International Library of Ethics, Law, and the New Medicine

Volume 69

The book series International Library of Ethics, Law and the New Medicine comprises volumes with an international and interdisciplinary focus. The aim of the Series is to publish books on foundational issues in (bio) ethics, law, international health care and medicine. The 69 volumes that have already appeared in this series address aspects of aging, mental health, AIDS, preventive medicine, bioethics and many other current topics. This Series was conceived against the background of increasing globalization and interdependency of the world's cultures and governments, with mutual influencing occurring throughout the world in all fields, most surely in health care and its delivery. By means of this Series we aim to contribute and cooperate to meet the challenge of our time: how to aim human technology to good human ends, how to deal with changed values in the areas of religion, society, culture and the self-definition of human persons, and how to formulate a new way of thinking, a new ethic. We welcome book proposals representing the broad interest of the interdisciplinary and international focus of the series. We especially welcome proposals that address aspects of 'new medicine', meaning advances in research and clinical health care, with an emphasis on those interventions and alterations that force us to re-examine foundational issues.

More information about this series at http://www.springer.com/series/6224

Kristien Hens • Daniela Cutas
Dorothee Horstkötter
Editors

Parental Responsibility in the Context of Neuroscience and Genetics

Springer

Editors
Kristien Hens
Department of Philosophy
University of Antwerp
Antwerp, Belgium

Dorothee Horstkötter
Department of Health, Ethics and Society
School of Mental Health and Neuroscience (MHeNS)
University of Maastricht
Maastricht, The Netherlands

Daniela Cutas
Department of Historical, Philosophical and Religious Studies
Umeå University
Umeå, Sweden

Department of Philosopy, Linguistics and Theory of Science
University of Gothenburg
Gothenburg, Sweden

ISSN 1567-8008 ISSN 2351-955X (electronic)
International Library of Ethics, Law, and the New Medicine
ISBN 978-3-319-42832-1 ISBN 978-3-319-42834-5 (eBook)
DOI 10.1007/978-3-319-42834-5

Library of Congress Control Number: 2016957395

Printed on acid-free paper

This Springer imprint is published by Springer Nature
The registered company is Springer International Publishing AG Switzerland

Preface and Acknowledgments

New technological developments and scientific advances raise increasingly complex questions about the nature and extent of parental responsibility. Findings in the field of genetics and embryology, as well as new understandings of how our brains work and develop, complicate these matters even more. This creates a need to reflect on the ways in which parental responsibility can or should be operationalised in the face of these new challenges – a task we undertake in this volume.

The idea behind this volume arose over a lunchtime discussion between the editors, then colleagues at Maastricht University, in the spring of 2013. Having discovered a common thread in our respective research projects, we organised a symposium on parental responsibility later that year. The Centre for Society and the Life Sciences in the Netherlands supported this initiative, and the symposium was a great opportunity for us to learn more about how parental responsibility is reflected in each of our own research areas. It was this symposium that led us ultimately to the publication of the present volume.

We wish to thank the contributors for their work and their patience in preparing this volume. We also thank Rebecca Bennett, Inge Liebaers and David Shaw for reviewing some of the chapters and the participants at the symposium for their helpful comments.

Antwerp, Belgium — Kristien Hens
Umeå and Gothenburg, Sweden — Daniela Cutas
Maastricht, The Netherlands — Dorothee Horstkötter

Contents

Contributors

Gunnar Björnsson Umeå University, Umeå, Sweden

Anna M.T. Bosman Faculty of Social Sciences, Behavioural Science Institute, Radboud University Nijmegen, Nijmegen, The Netherlands

Bengt Brülde Department of Philosophy, Linguistics and Theory of Science, University of Gothenburg, Gothenburg, Sweden

Daniela Cutas Department of Historical, Philosophical and Religious Studies, Umeå University, Umeå, Sweden

Department of Philosophy, Linguistics and Theory of Science, University of Gothenburg, Gothenburg, Sweden

Wybo Dondorp Department of Health, Ethics & Society, School for Public Health and Primary Care (CAPHRI), Maastricht University, Maastricht, The Netherlands

Kathleen Emmery Knowledge Center, Higher Institute for Family Studies, University College Odisee, Brussels, Belgium

Ademola Kazeem Fayemi Department of Philosophy, University of Lagos, Lagos, Nigeria

Frans J.M. Feron Bioethics Institute Ghent, Department of Social Medicine, School for Public Health and Primary Care (CAPHRI), Maastricht University, Maastricht, The Netherlands

Farah Focquaert Department of Philosophy and Moral Sciences, Ghent University, Ghent, Belgium

Simona Giordano School of Law, The University of Manchester, Manchester, UK

Francisco Güell Pelayo Mind-Brain Group, Institute for Culture and Society (ICS), Universidad de Navarra, Edificio Bibliotecas (ICS) – Campus Universitario, Pamplona, Spain

Kristien Hens Department of Philosophy, University of Antwerp, Antwerp, Belgium

Peter Herissone-Kelly School of Humanities and Social Sciences, University of Central Lancashire, Preston, UK

Dorothee Horstkötter Department of Health, Ethics and Society, School of Mental Health and Neuroscience (MHeNS), University of Maastricht, Maastricht, The Netherlands

Sylvia Hübel Knowledge Center, Higher Institute for Family Studies, University College Odisee, Brussels, Belgium

Adelheid Rigo Higher Institute for Family Studies, University College Odisee, Brussels, Belgium

Elena V. Syurina Athena Institute for Research on Innovation and Communication in Health and Life Sciences, Faculty of Earth and Life Sciences, Vrije Universiteit, Amsterdam, The Netherlands

Department of Health, Ethics and Society, School for Public Health and Primary Care (CAPHRI), Maastricht University, Maastricht, The Netherlands

Hans Van Crombrugge Higher Institute for Family Studies, University College Odisee, Brussels, Belgium

Guido de Wert Department of Health, Ethics & Society, School for Public Health and Primary Care (CAPHRI), Maastricht University, Maastricht, The Netherlands

Biographies

Gunnar Björnsson is professor of philosophy at Umeå University, Sweden, where he is the principal investigator of the 'Responsibility in Complex Systems' project, which looks at issues of individual, shared and collective moral responsibility and relations between these. He is also the coordinator of the Moral Responsibility Research Initiative at the University of Gothenburg, which brings together researchers with applied and foundational interests in issues of moral responsibility. He has published on a wide variety of topics, including causation and conditionals, but his research focuses primarily on issues in meta-ethics and moral responsibility and he has been published in, e.g., *Mind*, *Ethics*, *Philosophy and Phenomenological Research*, *Noûs*, *Philosophical Studies* and *Australasian Journal of Philosophy*.

Anna M.T. Bosman is professor at the Department of Special Education and the Behavioural Science Institute of Radboud University, Nijmegen, the Netherlands. Her research involves effective instruction of reading and spelling in dyslexia as well as the interaction between support staff and children or young adults with mild intellectual disabilities and behavioural problems. She currently teaches the master course 'Complex systems theory'. An important recent publication related to the contribution in this volume is Bosman et al.'s (2013) 'From the Role of Context to the Measurement Problem: The Dutch Connection Pays Tribute to Guy Van Orden' (*Ecological Psychology, 25*, 240–247).

Bengt Brülde is professor of practical philosophy at the University of Gothenburg, Sweden. His current work focuses on climate ethics, the philosophy of love and friendship and collective and shared responsibility (he is a participant in the 'Responsibility in Complex Systems' project, funded by the Bank of Sweden Tercentenary Foundation). He has published on a wide variety of topics, including well-being, the concepts of health and disorder, the ethics of happiness and suffering, the meaningful life, bioethics, business ethics, public health ethics, global justice and enhancement. He is a member of the editorial board of the *Journal of Happiness Studies* and was until recently coeditor of philosophy for the same journal.

Daniela Cutas is associate professor of practical philosophy at Umeå University as well as at the University of Gothenburg, Sweden. She is the principal investigator of the 'Close Personal Relationships, Children and the Family: Ethical and Political Analysis Against a Changing Background' project, funded by the Swedish Research Council. Her current work focuses on the ethics of close personal relationships, reproduction and parenting and more broadly on bioethics and research ethics. She has published in these research areas in journals such as *Bioethics*, *Cambridge Quarterly of Healthcare Ethics*, *Human Fertility*, *Reproductive BioMedicine Online*, *Journal of Medical Ethics*, *Health Care Analysis*, *Science and Engineering Ethics* and *Hypatia*. She is a coeditor of the collective volume *Families: Beyond the Nuclear Ideal* (2012) (with Sarah Chan).

Guido de Wert is professor of biomedical ethics at Maastricht University. He is head of the Department of Health, Ethics and Society. His research interests concern the ethics of reproductive genomics and genetic testing. He is a member of the Health Council of the Netherlands.

Wybo Dondorp is associate professor of biomedical ethics at Maastricht University. He has previously worked as an ethicist for the Health Council of the Netherlands. His research interests concern the ethics of medically assisted reproduction and of screening policies. He is a member of the Committee on Late Abortion of the Dutch Society of Obstetrics and Gynaecology.

Kathleen Emmery is coordinator at the Knowledge Center of the Higher Institute for Family Sciences (University College Odisee, Brussels). She studied criminology (master) at the University of Ghent. Her main research topics are family policies and relationship support in Flanders.

Ademola Kazeem Fayemi is lecturer at Lagos State University, Nigeria, where he teaches classes on ethics, African philosophy and sociopolitical philosophy. He is a fellow of the Erasmus Mundus European Master programme in bioethics, Catholic University of Leuven. He has published widely in international and national journals of philosophy, bioethics and African studies. His interests are in Yoruba philosophy, bioethics and research ethics in Africa.

Frans J.M. Feron has a background in medicine, is trained as a medical doctor (MD) and officially registered as a medical specialist in community health and social medicine in the field of child and youth health care. In addition, he is professor in social medicine, in particular child and adolescent health, and chair of the Department of Social Medicine at Maastricht University. He has 35 years of expertise in preventive paediatric primary care for children and adolescents at the Youth Health Care Division of the Regional Public Health Service (GGD), South Limburg. Alongside his clinical work, he developed a keen interest in neurodevelopmental issues in children, in child mental health and in early detection and treatment of psychopathology. He completed his PhD in medicine based on studies on neurodevelopmental issues in children. His current research activities focus on health, growth and development of children and adolescents, in particular mental health and behavioural problems, and specifically (neuro)developmental issues in children and adolescents.

Farah Focquaert obtained her PhD in 2007 from the University of Ghent in Belgium. She is a visiting professor of public health ethics at HoWest College, Oostende, and a research fellow at the Bioethics Institute Ghent, University of Ghent. Her work is predominantly situated in the fields of neuroethics and philosophy of free will. Her current research focuses on the ethics of moral enhancement, the ethics of neuromodulation for treatment and enhancement (e.g. questions related to narrative identity, regulation, assent/consent and criminal behaviour) and the implications of the philosophy and neuroscience of free will for the criminal justice system. She was a visiting research fellow at the Center for Cognitive Neuroscience, Dartmouth College, and a visiting scholar at the Department of Criminology and the Center for Neuroscience and Society, University of Pennsylvania. She is the chair of the Ethics Committee at the Forensic Psychiatric Center in Ghent, a research fellow of the Research Foundation Flanders and a member of the Moral Brain research group.

Simona Giordano is reader in bioethics at the Centre for Social Ethics and Policy, University of Manchester. She has been researching on nonconforming gender identity, and she has published *Children with Gender Identity Disorder* (Routledge, 2012), as well as many articles on the topic. She is regularly involved with charities supporting trans-acceptance in the UK and has worked for the amendment of clinical practice relating to minors with gender dysphoria in the UK and abroad.

Francisco Güell Pelayo is coordinator of the Mind and Brain Group at the Institute for Culture and Society at the University of Navarra (UNAV), Spain. His research interests include epigenetics, embryo development, bioethics and addiction. He has completed extended research stays at King's College, London, Georgetown University and the University of Stanford. He has published one book about the epigenetics of embryo development and over 15 articles in the past three years. He has attended over 50 international congresses and has been an invited speaker at the Kennedy Institute of Bioethics, the Center for Clinical Bioethics of Georgetown University, the School of Health and Social Care at Oxford Brookes University and Universidad Austral, Argentina. He is associate professor of the Philosophy Department at UNAV and invited professor on the master of bioethics course at Rey Juan Carlos University, Madrid.

Kristien Hens is a bioethicist whose work focuses on ethical issues related to genetics, reproductive medicine, feminism, disability studies and neurodiversity (autism, ADHD). She is currently working as a postdoctoral researcher at the Department of Philosophy of the University of Antwerp investigating the ethical issues related to biological explanations of autism. She is also a research fellow at the Centre for

Biomedical Ethics and Law of the KU Leuven. Before that she worked as post-doctoral researcher on the ethics of embryo selection, at Maastricht University, and as a PhD student at the KU Leuven on the ethics of the use of stored tissue samples from children for research. She has published in several medical and ethics journals, including *Journal of Medical Ethics*, *Human Reproduction* and *European Journal of Human Genetics*.

Peter Herissone-Kelly is a philosopher who has published widely in bioethics, with papers and book chapters on the ethics of new reproductive technologies, the desirability (or otherwise) of genetic enhancement and the theoretical foundations of bioethical inquiry. He is also coeditor of a number of edited collections on bioethics. He is senior lecturer in philosophy at the University of Central Lancashire, UK, where he teaches classes on, amongst other things, philosophy of language, metaphysics, epistemology and meta-ethics. He holds degrees from the Universities of Bolton, Oxford and Central Lancashire.

Dorothee Horstkötter is assistant professor at the Department of Health, Ethics and Society, School of Mental Health and Neuroscience (MHeNS), at Maastricht University in the Netherlands. She is trained as a philosopher, biologist and educational scientist. She obtained her PhD at Radboud University, Nijmegen, the Netherlands, with a thesis in philosophy entitled 'Self-Control Revisited: Varieties of Normative Agency'. She is the principal investigator of the project 'Parents who are alcoholics, Towards a normative framework of integrative care and responsible parenting interventions' funded by the Netherlands Organisation of Health Research and Development. Her research covers neuroethics, bioethics, public health ethics, medical humanities and philosophy of action with a special focus on (child) mental health issues. Her publications have appeared in *Bioethics*, *BioSocieties*, *Theory & Psychology*, *International Journal of Law and Psychiatry* and *The American Journal of Bioethics Neuroscience*, as well as in various Dutch and German professional journals.

Sylvia Hübel is a researcher in the Department of Theological and Comparative Ethics at the KU Leuven in Belgium. She holds an MA degree in family and sexuality studies and theology. She worked as a researcher in the framework of this practice-oriented project on prenatal diagnosis at the Higher Institute for Family Sciences (University College Odisee of Brussels) between 2011 and 2013. Her research interests include medical humanities, bioethics, women's studies, family and sexuality sciences.

Adelheid Rigo is lecturer and researcher in psychology and bioethics at the Higher Institute for Family Sciences of University College Odisee (Brussels). She studied philosophy (master, PhD), moral sciences (bachelor) and clinical psychology (master) at the Free University of Brussels (VUB). Her main research topics concern the ethical aspects of preimplantation genetic diagnosis with HLA typing, ethics and genetics and prenatal diagnosis (e.g. NIPT). She worked on several scientific projects in this area both as researcher and supervisor. The latest supervised project 'Prenatal Screening and Diagnosis: A Moral Imperative?' is a collaboration between the Odisee and the VUB.

Elena V. Syurina is a researcher at the Athena Institute for Research on Innovation and Communication in Health and Life Sciences, Faculty of Earth and Life Sciences, Vrije Universiteit, Amsterdam, the Netherlands and assistant professor at the Department of Health, Ethics and Society, School of Public Health and Primary Care (CAPHRI) Maastricht University, the Netherlands. She has a mixed background including public health, public policy analysis, child and youth health care and neuropsychology. Her present research revolves around the possibilities of changing the existing primary health system to make it more personalised. Another special interest is mental health of children and how it can be improved. An important publication related to the chapter in the present volume is "What about FH of my child?" Parents' opinion on family history collection in preventive primary pediatric care, by Syurina et al., published in *Personalized Medicine*.

Hans Van Crombrugge is senior lecturer in educational theory and family pedagogy at the Higher Institute for Family Sciences of University College Odisee (Brussels). He studied educational sciences at the KU Leuven. He has previously been a junior researcher at the Faculty of Psychology and Educational Sciences of the KU Leuven, director of the educational project 'Open Year' of the KU Leuven and lecturer of foundations of educational theory and family pedagogy at the University of Ghent. His research and publications deal with theory and history of educational thought in general. Some more specific research topics are the ethics of parent-child relationships, concepts of kinship, liberal education and plurality of worldviews, history and theory of Christian and Islamic philosophy of education and family life education.

Abbreviations

ADHD	Attention Deficit Hyperactivity Disorder
APA	American Psychiatric Association
ART	Assisted Reproductive Technology
ASD	Autism Spectrum Disorder
BOPZ	Bijzondere Opnemingen in Psychiatrische Ziekenhuizen (Psychiatric Hospitals Compulsory Admissions Act)
CBCL	Child Behaviour Checklist
CEOAE	Click-evoked otoacoustic emissions
CP	Cerebral palsy
CYHC	Child and Youth Health Care
DBS	Deep brain stimulation
DNA	Desoxyribonucleic acid
DSM	Diagnostic and Statistical Manual of Mental Disorders
FAS	Foetal alcohol syndrome
FASD	Foetal alcohol spectrum disorders
FDA	Food and Drug Administration
GP	General practitioner
HCP	Health Care Professional
ICSI	Intracytoplasmic sperm injection
IDRPD	International Declaration on the Rights of Persons with Disabilities
IVF	In vitro fertilisation
MBLC	Molecular Biology Laboratory Centre
MCC	Mother and Child Clinic
NAS	Neonatal abstinence syndrome
NGO	Non-governmental organisation
NIPT	Non-invasive prenatal testing
NNPC	Nigerian National Petroleum Corporation
OAE	Otoacoustic emissions
OCD	Obsessive-compulsive disorder
PD	Parkinson's disease

PDD-NOS	Pervasive developmental disorder – not otherwise specified
PFC	Prefrontal cortex
PGD	Preimplantation genetic diagnosis
PND	Prenatal diagnosis
PNS	Prenatal screening
PNT	Prenatal testing
PPB	Principle of procreative beneficence
RNA	Ribonucleic acid
RPHS	Regional Public Health Service
SCA	Sickle-cell anaemia
SDQ	Strengths and Difficulties Questionnaire
SRY	Sex-determining region of Y (chromosome)
TDF	Testis development factor
TS	Tourette syndrome
WHO	World Health Organization
WISC	Wechsler Intelligence Scale for Children
WMA	World Medical Association

Chapter 1
Parental Responsibility: A Moving Target

Kristien Hens, Daniela Cutas, and Dorothee Horstkötter

Abstract Beliefs about the moral status of children have changed significantly in recent decades in the Western world. At the same time, knowledge about likely consequences for children of individual, parental, and societal choices has grown, as has the array of choices that (prospective) parents may have at their disposal. The intersection between these beliefs, this new knowledge, and these new choices has created a minefield of expectations from parents and a seemingly ever-expanding responsibility towards their children. Some of these new challenges have resulted from progress in genetics and neuroscience. It is these challenges that we focus on in this introduction and volume.

Keywords Parental responsibility • Genetics • Neuroscience • Childhood • Parenting

1.1 Background

Ideas of what constitutes parental responsibility are constantly changing, under the influence of social, cultural, as well as scientific developments. A major contributor to discussions on responsibility for children is the changing landscape in the last

K. Hens (✉)
Department of Philosophy, University of Antwerp,
Stadscampus, Grote Kauwenberg 18, 2000 Antwerp, Belgium
e-mail: kristien.hens@uantwerpen.be

D. Cutas
Department of Historical, Philosophical and Religious Studies, Umeå University,
SE-901 87 Umeå, Sweden

Department of Philosophy, Linguistics and Theory of Science, University of Gothenburg,
Gothenburg, Sweden
e-mail: daniela.cutas@umu.se

D. Horstkötter
Department of Health, Ethics and Society, School of Mental Health and Neuroscience
(MHeNS), University of Maastricht, Postbox 616, 6200 MD, Maastricht, The Netherlands
e-mail: d.horstkoetter@maastrichtuniversity.nl

K. Hens et al. (eds.), *Parental Responsibility in the Context of Neuroscience and Genetics*, International Library of Ethics, Law, and the New Medicine 69,
DOI 10.1007/978-3-319-42834-5_1

century of shared perceptions of the moral status of children and their place in the family and in society. It is a relatively recent development that, at least in Western cultures, children are seen as bearers of rights as well as vulnerable members of our societies, in need of extra protection (Archard 2004) – sometimes from their own parents. From being considered cheap and expendable work force, little more than their parents' property, children have increasingly obtained legal and moral standing. In this process of progression from parent-centred accounts of parental rights towards ones in which children's rights are at least as important (Gheaus 2012), we are now leaving the familiar realm of parental duties as requiring not much more than that parents provide their children with a home, feed them, bring them to the doctor in case they fall ill, and send them to school. Today's expectations go much further than that, and social norms of what parents should do and what children are entitled to have fundamentally changed over the past decades. Parents are now held responsible for their children's cognitive and social-emotional development. Many parents feel obliged to introduce their children to various sports, have them learn how to play music instruments and one or two foreign languages from an early age. They are also being held responsible for the healthy physical development of their children, such as preventing them from becoming obese (Holm 2008). They should provide what is considered healthy foods, and are supposed to make use of educational interventions of various kinds that support children's emotional wellbeing. This does not only hold for what parents should do to improve their children's chances in life, but it holds likewise for their own behaviour and the role model that they have to offer. It has, for example, been argued that parents should not smoke in the presence of children and preferably should quit smoking altogether (Brennan and White 2007).

A much discussed aspect of parental responsibility is that of making medical decisions on behalf of children and in their *best interest* – and how this translates into practice. It is often not clear what children's best interest *is*, nor how a present interest balances against interests that a child might develop in the future. A major aspect of the discussion of what constitutes parental responsibility has to do with the appreciation of childhood. Is childhood only valuable as a transitory stage towards adulthood? Are the choices that a child may have as an adult in need of protection, even if this would mean an infringement of her current choices or wellbeing (Brennan 2014)? For example, if a child is at risk of losing her fertility, should parents safeguard her reproductive potential in case she wants to become a genetic parent later in life (Cutas in this volume)? If children are, so to speak, adults *in the making* [or *unfinished adults* (Gheaus 2015)] then perhaps their parents *do* have to prioritise at least some of the interests that children might have as adults over current interests. It is tempting to see a child's life in a historical perspective, but it is not clear that by choosing to safeguard an interest that the child does not yet have we are in fact working towards her best interests.

Moreover, respect for an adult person is often framed in terms of her autonomy, and only in exceptional cases can respect for adult autonomy be overridden by considerations of wellbeing. A child, however, is in a gradual developmental process towards autonomy and learns how to be an autonomous person by exercising her autonomy in a controlled environment. Hence, the autonomy of children, if any, can be restricted if such a restriction is deemed to be in her best interest (Matthews

1996). For example, Blustein argues that parents have a duty to respect a child's own desires in matters that are not critical to protecting the child's interests, and if they are not likely to impede the child's development (Blustein 1982). But this suggests that a child's best interest supersedes all other considerations. Often the process of medical decision-making on behalf of children goes smoothly: parents and health professionals agree on the kind of medical problem the child has and on how her condition should be best dealt with. However, any significant disagreement between parents and children, or parents and health professionals, can lead to open disputes that require a resolve of what the child's interests are, how they should be met, and who should meet them. Perhaps most notoriously, such have been the cases of Jehovah's witnesses refusing life-saving blood transfers for their children (Woolley 2005) or the cases of male circumcision for cultural rather than medical reasons (Earp 2013).

Questions regarding the scope of parental responsibility become even more acute given current new scientific findings in *neuroscience*. Controversies regarding the status of mental health disorders in children and the question whether, and if so which, children should go through diagnostic processes, complicate considerations about parental responsibility. Under which conditions and when is it appropriate to consider children to be at-risk to develop such disorders or to diagnose them? How should potential preventive interventions or treatments be envisaged? To answer these questions it makes a relevant difference whether mental health issues are seen as some kind of a brain-disease or as essentially normative interpretations of behaviour (see Bosman in this volume). Even if we agree on the fact that children with behavioural problems or mental health issues should be helped and that it is in the first instance the responsibility of their parents to seek and organize this help, as Syurina and Feron argue in this volume, how exactly parents can fulfil this responsibility varies greatly.

Progress in *genetics* also raises new questions with regard to responsibilities for children. For example, before genes and their functions were discovered, inheriting traits that run in the family was considered a result of fate rather than of specific choice. Premature death because of sickle cell anaemia, for example, has been called the 'born-to-die-prematurely syndrome' in societies affected by it (see Fayemi in this volume). With the discovery of the double helix and the many mutations associated with diseases, at first, this concept of fate was still there. As we cannot change our genes, the passing on of defective genes was seen as beyond our control. However, we can now be forewarned about a host of genetic conditions. In some cases, as that of sickle cell anaemia, this knowledge can be life-saving, in others it can make life for the child and her family tremendously easier. For example, if parents and professionals know that a child has phenylketonuria (PKU), and thus cannot metabolise certain foods, they can make certain dietary choices rather than others and thereby avoid that children will come to develop serious brain damage and mental retardation.

The possibility of detecting defective genes, in vitro or prenatally, raises important questions for parental responsibility (see also Hübel et al. in this volume). Should we choose between embryos, and if yes, how? Should we choose the embryo most likely to have the best start in life (see Herissone-Kelly in this volume;

Savulescu 2001; Savulescu and Kahane 2009)? Is the hope to create a child with the best chances in life realistic – or is this the result of an unjustifiably optimistic reliance on the impact of genetic conditions on the good life? Do the potential risks that these techniques may entail override the benefits of selecting the healthiest embryo (see Güell in this volume)? Once we *can* know our own genetic make-up, or the one of our gametes or embryos, do we have a responsibility to acquire this knowledge and act on it? If a negative outcome for children can be attributed to lifestyle choices rather than genetic causes, this responsibility is perhaps more easily attributed. Fetal alcohol syndrome is a seminal example. It is easier to blame a pregnant woman for drinking alcohol during pregnancy than to accuse a parent who unknowingly passes on a damaging mutation to the next generation. Are women morally blameworthy if their children are born with fetal alcohol syndrome or develop neonatal abstinence syndrome, and if they are, can they legitimately be pressured or coerced to stop drinking or using drugs in order to protect their children before they are born (see Dondorp and de Wert in this volume)?

The rise in knowledge about epigenetics complicates such questions even further. For example, studies have demonstrated that a diet low in vitamin B12 during pregnancy can induce problems in offspring, but also in the offspring of that offspring (Suren et al. 2013, Hens in this volume). Is a woman who has had a diet low in B12 responsible that her child is prone to becoming obese? Is she responsible for any consequence that this might have on her grandchildren? If there is a small risk that a man who smoked as a boy will cause asthma in his child many years later because of this (Hens 2017), and his child does develop asthma, is the man responsible for this outcome? The discovery of such connections between what we until recently saw as benign and private behaviours and circumstances, and (sometimes significant) effects on children or even future generations, can be baffling. It can, at the same time, invite moral judgment into areas of people's lives in ways that we may not see as justified, and it can create a temptation for policy-makers to police private lives in the name of future interests, future goods, or of the public good. All these are very important questions and risks that have to be discussed, and these discussions have to include scientists, psychologists, ethicists, and anyone else who can contribute relevant aspects.

Throughout this volume, "her" and "she" will be used when referring to adults and children in general – regardless of their gender. Unless otherwise specified, the term "parents" will be used to denote all caregivers taking up parental roles – regardless of genetic or biological connections. It will be used to denote both fathers and mothers, again unless otherwise specified. We are aware that expectations from (prospective) mothers and fathers, respectively, as well as from men and women, can differ greatly. As Purdy aptly pointed out already two decades ago, women are seen as owing more to their children before they are even mothers than anyone else owes anybody else (Purdy 1996). Perhaps an illustration of this is a court case in the US in which a child who needed compatible bone marrow to survive was denied access to his biological father's contact details (Purdy 1996, Re George 1982). In this case, not only was the man not expected to make a small effort to save the life of the child he helped create, but it was deemed unacceptable to so much as ask him. While these inequalities are increasingly being questioned and analysed in recent years, and the parental roles of mothers and fathers are coming closer together (Collier and Sheldon 2008), we acknowledge that much work still remains to be done.

1.2 Current Debates

While questions of parental responsibilities are actualised in various areas of life, the authors in this volume focus on those that are raised by findings in genetics and neuroscience. We investigate both whether there are *new* dimensions of parental responsibility that did not exist before the rise of genetics and/or neuroscience, and potential changes to the content of parental responsibilities already established (see Björnsson and Brülde in this volume). In the context of genetics, the possibility has arisen that parents select, at least partly, the genetic profile of their future child by means of, for example, preimplantation genetic diagnosis (PGD). This raises completely new questions such as whether parents have a responsibility to make use of such a technique to either avoid serious genetic diseases in their children, or maybe to enhance their 'genetic fitness' in general. In the context of neuroscience, enormous insights have been gained on the process of brain development during childhood as well as on the effects of children's environment on that development. This raises the question of whether it is part of parents' responsibility to take insights on brain development and potential adverse or beneficial environmental effects into account when having and raising children. In other words, should parents help their children train their brains in order to safeguard or even increase individual flourishing (see Horstkötter in this volume) or avoid socially deprived neighbourhoods because of potentially harmful effects on their children's brains? One could also argue that parents have a responsibility to help their children flourish and avoid their suffering even if this involves surgical intervention on the children's brain, as is the case of deep brain stimulation (see Foquaert in this volume).

What is meant by wellbeing is still under debate. A specific context in which responsibility towards future children has been extensively discussed is that of embryo selection. This context is different from that of prenatal testing, as there the responsibility is towards a specific (future) child. Indeed, embryo selection involves choosing between embryos in vitro based on chromosomal or genetic characteristics. Therefore, the question here is 'what children should there be' rather than what is good for a specific child. The concepts of *beneficence* or *enhancement* are central to this discussion. A much discussed principle in this context is that of *procreative beneficence*. According to this principle, prospective parents should pick the embryo that is most likely to have the best start in life (Savulescu 2001; Savulescu and Kahane 2009). This principle is a comparative one, and it does not prescribe that if only suboptimal embryos are available, none should be selected. It is also compatible with the exercise of parents' reproductive autonomy: parents should not be prevented from making choices even if this means choosing an embryo with a certain condition and violating the principle (Savulescu 2002).

Some authors have defined thresholds that an embryo should meet in order to be eligible for transfer to the womb. A minimal threshold suggests that there is a duty only to discard those embryos that would develop into children whose life would be not worth living because it would be filled with suffering. Others define an acceptable outlook threshold or even a maximum threshold, stating that only embryos with a very good prospect should be selected (Harris 2001; Glover 2006). Commentators have questioned the idea that prospective parents have an obligation *to select the best children*, rather than *the best for a specific child* (Bennett 2009, 2014; Herissone-Kelly 2006 and in this volume; McDougall 2005; Parker 2007). Parfit has argued that there are responsibilities to make sure that the wellbeing of the children that are born is maximized, even when this means selecting one child to be born over another or waiting for the optimal circumstances in which to procreate (Parfit 1984). This is an ongoing discussion that teases out intuitions and expectations from and of (prospective) parents when contemplating reproduction and what they should do.

In the creation of this volume and the selection of contributing chapters, we bring together perspectives relevant for the general topic. Parents, as opposed to non-parents, are considered to hold a special relationship to children. This relationship is characterized by the care that parents provide, and are expected to provide, to their children and is particularistic, referring to the relationship of a parent to a specific child or children. In this volume, we do not presuppose a specific ethical framework, but instead discuss, from different perspectives, the questions, problems and subjects generated by recent research in genetics and neuroscience for the role of parents. For this reason, we chose to focus on the more general term of 'parental responsibility' rather than that of 'parental duties'. Duties are more specific and can be formulated after a normative framework on how to deal with a specific problem or challenge has been determined. By looking at parental responsibilities, we allow for a more broad investigation of what parents should do, or whether they should do anything special at all, in order to live up to the potential demands made visible by certain scientific developments. Another relevant concept in this regard is that of 'parental obligations'. Following Björnsson and Brülde's elaboration in this volume, we take obligations to be specific requirements that arise from normative responsibilities. This keeps intact the expression of 'parental responsibility' as being the more comprehensive one that is therefore apt to bring together the chapters that are forming this volume.

1.3 This Volume and Its Chapters

Findings in genetics, epigenetics and neuroscience raise important questions about parental responsibility. Should parents endeavour to make their children as *normal* as possible to avoid all problems that they may encounter if they don't fit in? Is self-control a desirable characteristic, and what does it mean to increase it (see Horstkötter in this volume)? If a neurological condition is embedded in the child's identity, should we still try to remove or treat it (see Hens in this volume)? How about children whose gender identity does not match stereotypical gender roles (see Giordano in this volume)? Does parental responsibility include preserving a child's fertility (see Cutas in this volume)? How is parental responsibility conceived in non-Western cultures (see Fayemi in this volume)? What do prospective parents themselves think about issues related to the wellbeing of their future children, in the context of prenatal screening (see Hübel in this volume)? The chapters address the various new – and sometimes not so new – questions that arise in the context of genetics and neuroscience on the content and extent of parental responsibility.

In Chap. 2, Gunnar Björnsson and Bengt Brülde provide a theoretical background to the idea of (parental) responsibility. That parents are responsible for the wellbeing of their children is a very common assumption in discussions of what (prospective) parents should do, however these discussions often are unfolded without a clear, explicit, and shared understanding of what responsibility even *is*. Björnsson and Brülde develop these theoretical clarifications in our volume by investigating the relationship between normative responsibilities and corresponding obligations and demands. They argue that normative responsibilities are constituted by normative requirements that the responsible agents *care* appropriately about how well things go in certain regards, and that obligations generally can be seen as straightforward upshots of requirements to care. They also investigate what might be the sources of parental responsibilities: these may include capacities and costs required for taking on the responsibility in question, retrospective and causal responsibility, promises or contracts, and certain social relationships.

In Chap. 3, Anna Bosman criticises the current focus on the Diagnostic and Statistical Manual of Mental Disorders (DSM) and the assumption that medical distinctions between illness and health or between normal and pathological conditions have an objective, *scientific* character. As such, this chapter tackles fundamental issues of how we even define diseases and disorders, which need to be discussed before the scope of parental responsibility can be addressed in practice. Relying on the work of George Canguilhem, she develops a conceptual analysis of disorder. She argues that medicine and psychiatry are not sciences in the same sense as physics and chemistry, because the difference between health and illness, as between normal and abnormal, is always a normative one. It is values, not objective numbers, that determine whether a certain blood level is acceptable or not. Such evaluations cannot focus on individual organisms as such; instead they take into account the relationship between the organism and its environment. Hence a disorder, or as Bosman prefers to term it – a different order – cannot be located in any one individual, instead it is a result of the relationship between the organism and its environment. This understanding about the very meaning of disorder or pathology leads her to some interesting conclusions regarding the responsibilities of parents and other caregivers. They do not need to determine whether someone is different, but whether she suffers. This, however, requires that parents, caretakers and clinicians always reflect upon the norms embedded in current diagnoses and interventions.

In Chap. 4, Elena Syurina and Frans Feron reflect on the current system of child and youth health care (CYHC). They argue that a different theoretical framework is necessary to organize care such that children with developmental problems can be helped in a more timely and efficient fashion. Currently, CYHC support and monitor the development of children from birth to adolescence with a special focus on behavioural and psychosocial issues. Recent findings in both genetics and neuroscience on brain development and on the origin of behavioural disturbances show that problems develop step by step and over considerable periods of time. The authors' main concern is that currently care is provided to children only when they have approached the final and most serious stage of full-blown disorders. However, given that children gradually 'grow into deficit', and given that remission can be achieved more easily during preclinical phases, this approach should be fundamentally changed. CYHC should no longer focus on full-fledged mental health diagnoses, but should be involved already in early preclinical phases, provide early preventive interventions and safeguard access to care for all children who are burdened with difficulties in their psycho-social development.

In Chap. 5, Dorothee Horstkötter investigates the meaning of the concept of self-control, the relevance of childhood self-control for wellbeing and any repercussions these might have on the responsibilities of parents to raise self-controlled children. As she shows in a brief overview of up-to-date research in social psychology, neuroscience, and analytic philosophy, self-control has mainly been linked to willpower and taken to refer to the successful overriding of any inadequate spontaneous responses for the sake of desirable yet distant goals. A critical analysis of current approaches, however, suggests that what has been investigated is 'controlled' behaviour, while the prefix 'self' is hardly given explicit attention. Horstkötter develops a comprehensive understanding of the term that does justice to both aspects. This in turn gives rise to two different sets of educational goals that have separate implications of what it even means to raise self-controlled children. Raising self-controlled children can entail, firstly, to teach them psychological or brain-training strategies to overcome temptation. Secondly, it can imply that parents should support children to develop a self that sets its own goals, reflects on these goals, and considers them as reasons for action. In this sense, while being informative, current neuroscience and social psychology miss out on their aim to identify what is required to raise self-controlled children, because they cannot determine the value children put on any distant goals or spontaneous desires.

Sometimes no established medical treatment can help relieve children's suffering. Farah Focquaert discusses in Chap. 6 the case of paediatric deep brain stimulation (DBS). DBS involves brain surgery and direct stimulation of the brain via electrodes. In the case of Parkinson's disease and essential tremor, this treatment has proved effective in adults. In children, DBS has been used most commonly for dystonia, yet at present it is considered investigational treatment for all paediatric conditions. Paediatric DBS hence is applied only in the framework of therapeutic medical research. Focquaert addresses the ethics of paediatric decision-making and participation in this context. She focuses on two ethically salient issues. First, typically, it is parents who make decisions on behalf of their children and who give, or do not give, their informed consent for treatment. Focquaert, however, argues that shared-decision making that involves the child patient, the parents and the medical team provides the strongest safeguards for the child's best interest. Second, it has been frequently argued that one should first have confirmed successful treatment outcomes in adults before applying the same measures to children. Under certain conditions, however, this rationale should not be followed and instead investigational paediatric DBS might be justified even though no evidence from adults is available.

In Chap. 7, Kristien Hens investigates how epigenetics complicates current discussions on the responsibility of the pregnant woman towards her future child. Epigenetics is a discipline that aims to understand how environmental factors influence organisms on a molecular level and identify how these factors can affect the expression of genes. Epigenetic influences may be heritable and reversible, challenging current assumptions about responsibility in the ethics of genetics. In neurology it is believed that epigenetics partly explains the development of neurological conditions and plays an important role in synaptic plasticity. As many epigenetic changes happen in utero, maternal behaviour may affect brain development. After first discussing new questions raised by epigenetics, Hens then uses the examples of autism and high intelligence ("giftedness"), to investigate the distinction between prevention and enhancement. She describes how some autistic people consider autism a difference rather than a disability, which should be accommodated for rather than cured. Moreover, high intelligence, which is often used as an example of a desirable trait, can lead to social and educational challenges as well. The fact that neurological difference may be considered an identity rather than an affliction is relevant to the discussion about maternal responsibility and epigenetics. If neurodifference is an identity with a value on its own then such responsibility would not entail trying to prevent or cure it.

In Chap. 8, Wybo Dondorp and Guido de Wert discuss the case of pregnant addicted women and the responsibilities that they may have towards their future children. They are particularly concerned about cases in which pregnant women are unwilling or unable to stop their drug use and thereby directly endanger the health and wellbeing of the child. While prenatal child protection is a morally important good, they argue, it is not enough to justify pressure and coercion against pregnant women. Given that any strong pressure (such as requests for legal supervision) and coercion (such as forced hospitalization of pregnant women to avert danger from the child) interfere with women's rights to self-determination, further criteria must be met. In addition, the harm to be prevented must be plausible, and the measures imposed must fulfil criteria of effectiveness, proportionality and subsidiarity.

In Chap. 9, Simona Giordano reviews the current discussion on gender issues and writes about parental and social responsibilities for children's gender identity development. She presents recent research on sex and gender identity formation and shows that at least some gender differences are not socially constructed and are expressed already before birth. Moreover, at least in some cases gender identification is congruent with sex differences that are not as immediately evident or testable as are genitals or sex chromosomes. Hence, Giordano argues, although biology may play a role in determining which gender a child identifies with, there is no fixed set of biological markers that can allow us to determine whether an individual is a female or a male. The distinction between sex and gender becomes more complex, as even the biological concept of sex has many different gradations. Moreover, the gender a person identifies with is not always stable across her lifespan. Giordano proposes that sex and gender are treated as broad approximations along a rich spectrum of possibilities. Parents as well as society in general have a responsibility to avoid gender stereotypes and not to insist on binary distinctions.

In Chap. 10, Peter Herissone-Kelly argues that prospective parents are not bound in their reproductive decision-making by the principle of procreative beneficence (PPB). According to the PPB, reproducers have an obligation to choose the embryo that is most likely to become the child that will lead the best life. Herissone-Kelly argues that the considerations that constitute the PPB reflect an *external perspective*, one that considers which future child, from possible children, is likely to have the best life. Such a perspective may be appropriate for policy makers. The appropriate perspective for prospective parents is, however, an *internal* one: parents consider what it is like for their specific future child to live a certain life. The sorts of considerations that underlie the principle of procreative beneficence do not constitute sufficient reasons to imply an obligation for prospective parents. Hence, there can be no requirement for prospective parents to be moved by those considerations.

Francisco Güell takes another route in Chap. 11 to challenging the PPB. Instead of examining, like Herissone-Kelly, whether parents are bound by the principle, he depicts a complex picture of the array of risks and black boxes that are actualised by the use of IVF, which is a necessary step to put the PPB into practice. On that basis, he casts doubts onto whether by acting on the PPB we really are *avoiding* rather than *creating* risks. Not talking about the risks involved in IVF, while emphasising risks that parents could actualise in their daily life (by smoking, alcohol intake or stress), creates an arbitrary distinction between the two sources of risk. In the current situation, the latter of these situations is presented as heavily loaded with meaning from a responsibility perspective, while the former is seen as responsibility free. Güell makes a case for equipping prospective parents with adequate information to enable them to make reproductive choices – and this information must include an honest evaluation of the risks involved in making recourse to IVF, particularly if that would not be for reasons of infertility but for the aim of procreative 'benefit'.

In Chap. 12, Daniela Cutas discusses the question of whether parents have a responsibility to take active steps to *rescue* their children's fertility. Starting with the case of treatments with a high risk of rendering children infertile, she examines the case for fertility preservation, as well as some of its possible implications: do only children who run an immediate risk of losing their fertility have a right that their parents (or someone else) rescue their fertility, or do others as well, and on what grounds? If (some) children have a right to fertility preservation, why should it be up to the parents to determine whether this right should be exercised? Can we draw a distinction between fair and unfair, deserved and undeserved capacity to reproduce – and use it in practice? By discussing these questions and more, Cutas tests intuitions and arguments and unfolds some of the complexities of the idea that parents have responsibilities regarding their children's reproductive capacities.

Illustrating the tension between deeply held beliefs and understandings of genetic conditions and the explanations and possibilities offered by modern medicine and genetics, Ademola Fayemi explores in Chap. 13 the example of sickle cell anaemia (SCA) in Yoruba culture in Nigeria. In this culture, responsibility for children is shared between members of the extended families in which children are born – and not limited to the children's parents (according to a Yoruba proverb, "children are biologically born by two eyes (…) but collectively nurtured by more than a thousand eyes"). This helps lift some of the burdens of coping with difficult situations in children's lives, but at the same time sustains age-old explanations and treatments that discourage uptake of information and support from modern medicine and technology. This case is a powerful and telling example of the need to find the best solutions for children, their parents, their families, and ultimately entire communities, in a way that profits at the same time from the goods in a culture and the goods in science and technology.

As is by now apparent from the more theoretical chapters in this volume, the issue of whether prospective parents have a duty to select the healthiest possible children is controversial and much discussed in recent years. In Chap. 14, Sylvia Hübel and colleagues present an empirical study in which they gained insight into views and attitudes regarding prenatal diagnosis and parental autonomy among health care professionals and parents who had recently undergone prenatal testing. They found that many (prospective) parents reject the idea of parental responsibility as having to select the healthiest possible children. Instead, parents have internalised the responsibility to respect lifestyle recommendations during pregnancy and to give the best possible care after birth. Hübel and colleagues also found that health care professionals thought that the main task of counselling in the case of prenatal testing was to provide prospective parents with help in making their own informed choices. However, at the same time, health care professionals also reported ethical dilemmas, such as what to do if prospective parents wanted testing or termination for minor anomalies, or if they did not want testing even though they had a high risk of transmitting a severe disease. In sum, professionals subscribe to the ideal of non-directive counselling, although parents as well as professionals themselves doubt the feasibility and desirability of adhering to this ideal in all cases.

References

Archard, D. 2004. *Children: Rights and childhood*. London: Psychology Press.

Bennett, R. 2009. The fallacy of the principle of procreative beneficence. *Bioethics* 23(5): 265–273.

Bennett, R. 2014. When intuition is not enough. Why the principle of procreative beneficence must work much harder to justify its eugenic vision. *Bioethics* 28(9): 447–455.

Blustein, J. 1982. *Parents and children: The ethics of the family*. New York: Oxford University Press.

Brennan, S. 2014. The goods of childhood and children's rights. In *Family-making, Contemporary ethical challenges*, ed. F. Baylis and C. McLeod, 29–45. Oxford: Oxford University Press.

Brennan, S., and A White. 2007. Responsibility and children's rights: The case for restricting parental smoking. In *Taking responsibility for children*, ed. Samantha Brennan and Robert Noggle, 97–112. Waterloo: Wilfrid Laurier University Press.
Collier, R., and S. Sheldon. 2008. *Fragmenting fatherhood. A socio-legal study*. Oxford: Hart Publishing.
Earp, Brian D. 2013. The ethics of infant male circumcision. *Journal of Medical Ethics* 39(7): 418–420.
Gheaus, A. 2012. The right to parent one's biological baby. *Journal of Political Philosophy* 20(4): 432–455.
Gheaus, A. 2015. Unfinished adults and defective children: On the nature and value of childhood. *Journal of Ethics and Social Philosophy* 9(1): 1–21.
Glover, J. 2006. *Choosing children: Genes, disability, and design*. London: Oxford University Press.
Harris, J. 2001. One principle and three fallacies of disability studies. *Journal of Medical Ethics* 27(6): 383–387.
Hens, K. 2017. The ethics of postponed fatherhood. *International Journal of Feminist Approaches to Bioethics* 10(1): 1.
Herissone-Kelly, P. 2006. Procreative beneficence and the prospective parent. *Journal of Medical Ethics* 32(3): 166–169.
Holm, S. 2008. Parental responsibility and obesity in children. *Public Health Ethics* 1(1): 21–29.
Matthews, G. 1996. *The philosophy of childhood*. Cambridge, MA: Harvard University Press.
McDougall, R. 2005. Acting parentally: An argument against sex selection. *Journal of Medical Ethics* 31(10): 601–605.
Parfit, D. 1984. *Reasons and persons*. Oxford: Oxford University Press.
Parker, M. 2007. The best possible child. *Journal of Medical Ethics* 33(5): 279–283.
Purdy, L. 1996. *Reproducing persons. Issues in feminist bioethics*. Ithaca: Cornell University Press.
Re George 1982. 630 S.W.2d 614.
Savulescu, J. 2001. Procreative beneficence: Why we should select the best children. *Bioethics* 15(5–6): 413–426.
Savulescu, J. 2002. Deaf lesbians, "designer disability", and the future of medicine. *British Medical Journal* 325: 771.
Savulescu, J., and G. Kahane. 2009. The moral obligation to create children with the best chance of the best life. *Bioethics* 23(5): 274–290.
Suren, P., C. Roth, M. Bresnahan, M. Haugen, M. Hornig, D. Hirtz, … , C. Stoltenberg. 2013. Association between maternal use of folic acid supplements and risk of Autism in children. *JAMA : The Journal of the American Medical Association* 309(6): 570–577.
Woolley, S. 2005. Children of Jehovah's witnesses and adolescent Jehovah's witnesses: What are their rights? *Archives of Disease in Childhood* 90(7): 715–719.

Chapter 2
Normative Responsibilities: Structure and Sources

Gunnar Björnsson and Bengt Brülde

Abstract Normative responsibilities have a central role in everyday moral thinking, largely because they are taken to ground requirements to act and react in certain ways. If parents are responsible for the wellbeing of their children, for example, this might mean that they are morally required to feed them, attend to their emotional needs, or make sure that someone else does. But normative responsibilities are not well understood as lists of requirements to act or react, for such requirements will depend on what options and information the agent has available. In the first part of the paper, we instead propose to understand normative responsibilities as requirements to care about what one is responsible for: about the wellbeing of one's child, about performing a certain action, or about playing the sort of role that one's profession requires. Such requirements, we argue, are just the sort of things that will give rise to requirements to act and react given the right context. In the second part, we survey and discuss a variety of considerations that might give rise to normative responsibilities: capacities and costs; retrospective and causal responsibility; benefits; promises, contracts and agreements; laws and norms; and roles and special relationships.

Keywords Responsibility • Normative responsibility • Prospective responsibility • Parental responsibility • Moral requirements • Moral obligations • Role responsibility

G. Björnsson (✉)
Umeå University, Umeå, Sweden
e-mail: gunnar.bjornsson@umu.se

B. Brülde
Department of Philosophy, Linguistics and Theory of Science, University of Gothenburg, Gothenburg, Sweden

K. Hens et al. (eds.), *Parental Responsibility in the Context of Neuroscience and Genetics*, International Library of Ethics, Law, and the New Medicine 69,
DOI 10.1007/978-3-319-42834-5_2

2.1 Introduction[1]

Attributions of what we shall call *normative responsibilities* play a central role in everyday moral thinking. It is commonly thought, for example, that parents are responsible for the wellbeing of their children, and that this has important normative consequences. Depending on context, it might mean that parents are morally required to bring their children to the doctor, feed them well, attend to their emotional needs, or to see to it that someone else does. Similarly, it is sometimes argued that countries that emit most greenhouse gases have most responsibility for preventing catastrophic climate change. This responsibility might imply that these countries are morally required to take necessary steps individually and jointly to implement a workable prevention plan, and to avoid steps that worsen the situation. More trivially, the grading of your student's essays might be your responsibility, as might making sure there is wine at tomorrow's picnic, and you might thus be required to see to it that essays are competently graded and suitable wine brought to the picnic.

Normative responsibility is distinct from two other relations called "responsibility": *retrospective moral responsibility*, i.e. the sort of relation an agent must stand in to decisions, actions, and outcomes in order to deserve blame or credit for them, and *causal responsibility*, which we have in mind when asking whether something caused something else ("is the infection responsible for these symptoms?").[2] Retrospective and causal responsibility might *give rise* to normative responsibilities—parents might be responsible for the wellbeing of their children because they are causally and retrospectively morally responsible for their existence. Still, normative responsibilities are themselves primarily prospective, and are often grounded in what can be done rather than in what has been done: it might be your responsibility to help the drowning child, not because you are retrospectively or causally responsible for its situation, but because you are the only one who can save the child.

Normative responsibility is also distinct from two properties of agents that go by "responsibility": the property of satisfying general conditions for being held responsible—for being blamed or credited—for one's actions (*capacity responsibility*), and the property of having a responsible character (*virtue responsibility*). People also talk about "taking" responsibility, having in mind acting or committing to act as is appropriate when one is retrospectively, causally, or normatively responsible for something ("He ought to take responsibility for their failure"; "you need to take responsibility for your role in all this"; "I take responsibility for setting up a

[1] This chapter has benefitted from a wide range of useful comments from participants at seminars at Umeå University and the University of Gothenburg, as well as from written comments by Kalle Grill, Jonas Nilsson, and the editors of this volume. Work on the chapter was funded by Stiftelsen Riksbankens Jubileumsfond.

[2] It is a matter of contention whether we should acknowledge other kinds of retrospective moral responsibility, distinct from the relation that grounds blameworthiness, and if so how many. See e.g. Smith 2012.

workable agreement"), and about deciding or acting responsibly, or in a responsible manner—a manner appropriate to the risks involved (*manner responsibility*).[3]

Though attributions of normative responsibilities are plentiful, such responsibilities have received surprisingly little philosophical attention compared to its normative relatives, *obligations* and *reasons*, and compared to retrospective responsibility. This chapter hopes to improve on this situation by taking on two main tasks. The first, attempted in Sect. 2.2, is to spell out the general structure of normative responsibility, in particular the relation between normative responsibilities and corresponding moral requirements and demands. We will suggest that normative responsibilities are constituted by normative requirements that the responsible agents *care* appropriately about how well things go in certain regards, and that moral requirements generally can be seen as straightforward upshots of requirements to care.

The second task, taken on in Sect. 2.3, is to provide an overview of prominent sources of normative responsibility and its distribution among agents. Why would the children's wellbeing be the parents' responsibility? Why not the neighbour's, or the state's, or everyone's? Here we discuss a range of possible sources, including agents' abilities, costs involved in taking on the responsibility in question, retrospective responsibility for the situation, promises or contracts, and certain social relationships.

The ambition of this chapter is to contribute to the present volume on parental responsibility by providing general perspectives on normative responsibilities and their sources. To clarify the relevance of the discussion to parental responsibility, we exemplify with cases involving parents and children, but to bring out the general nature of the conclusions we want to draw, other sorts of cases are also introduced.

2.2 The Structure of Normative Responsibilities

If one is normatively responsible for something—a child, the grading, the wine, the climate—one might be liable to be *held responsible* for certain states of it, such as the poor health of a child, the tardy grading, the lack of good wine, or the catastrophic climate, or for not taking reasonable measures to prevent these states. In the case of *legal* normative responsibilities, for example, the responsible party might be legally liable to fines, imprisonment, or other sanctions. Our focus here, however, is with *moral* normative responsibility. Those who fail to live up to such responsibilities are thereby liable to moral blame (cf. Miller 2001: 454).

[3]For an influential discussion of different meanings of "responsibility", see Hart 1968: ch. 9 (cf. Vincent 2011). For manner responsibility, see e.g. Munthe 2011: ch. 5, and discussions of a "duty of care" in tort law. What concerns us here is most closely identified with what Hart calls "role-responsibility". See also Robert E. Goodin's notion of "task responsibility" (Goodin 1995: ch. 5; Schmidtz and Goodin 1998: 150–54).

In what follows, we will talk about what one must or must not do in order to avoid moral blame. For ease of exposition, let us say that:

MORAL REQUIREMENTS: One is "morally required to φ" insofar as failure to φ (to act, omit to act, or to be in a certain way) makes one liable to moral blame to some degree for not φ-ing.[4]

Blame can be due for events or states of something that one is retrospectively morally responsible for, but also for actions, omissions, and certain attitudes or character traits (being uncaring, selfish, amoral). In addition, blame can come in different degrees of seriousness, depending on how far short one falls of what is required. Correspondingly, moral requirements, as defined here, have a variety of objects, and come in various strengths.

Given MORAL REQUIREMENTS, we can now say that *moral normative responsibilities* are responsibilities that give rise to moral requirements. (In what follows, when we talk about "responsibilities" and "requirements", we refer to *moral* responsibilities and requirements unless otherwise noted.)

Agents are often subject to requirements in virtue of their responsibilities, and it might be natural to think of responsibilities as constituted by lists of requirements to do various things, perhaps in particular to ensure certain outcomes and seek information relevant to those outcomes.[5] On further reflection, however, it is clear that such lists of requirements fail to capture what responsibilities are. The problem is that the requirements associated with a given responsibility seem highly context-dependent. For example, if parents are responsible for the health of a child, they are not thereby necessarily to blame for failure to prevent illness. They might not have been *able* to prevent the illness,[6] and even if they were, there are limits to the resources they were required to spend to eliminate small risks, as well as moral constraints on the means they might have employed to help their child (such as stealing medication from another child). Similarly, although parents might typically be required to pay attention to the child's development and consider risks and options, whether failures to do so are blameworthy will depend on opportunities for taking such action (parents might be separated from children for no fault of their own), on risks involved (some children and situations might require more attention), on resources available, and again on moral constraints (for example, as children grow, parental responsibility is increasingly constrained by the children's growing autonomy). In these ways, what results parents are required to achieve or prevent,

---

[4] Talk of moral requirements, as defined here, is arguably closely related to and perhaps equivalent to much ordinary talk of moral obligations, of what one ought to do, and of what it would be morally wrong not to do. However, uses of "wrong", "ought", "obligatory", or "required" in moral contexts are far from univocal, and there is legitimate talk of blameless wrongdoing and blameless failures to discharge obligations, or do what one ought or is required to. Defining a notion of *moral requirements* in terms of liability to blame (and later analyzing responsibilities in terms of such requirement) lets us avoid these ambiguities.

[5] See e.g. Goodin 1995: 82; Narveson 2007: 19.

[6] Generally, being required to φ seems to imply being able, in some sense, to φ, though the required kind of ability has been notoriously difficult to spell out.

what actions they are required to perform, and what they are required to pay attention to will depend on what risks and feasible and acceptable options there are. Specific requirements might also depend on the parents' degree of responsibility; one might think that one parent is more responsible for a child than the other parent, perhaps for being the one intentionally bringing about the pregnancy, or for promising to take most of the responsibility.

What might responsibilities be, then, such that they give rise to this highly context-dependent set of requirements? We propose a two-part explanation, first analysing responsibilities as *requirements to care* about the object of responsibility, and then explaining how such requirements to care give rise to other requirements. We look at these parts in Sects. 2.2.2 and 2.2.3, respectively. First, though, we need to say a little more about what it is to *care about* something, in the relevant sense.

2.2.1 Caring as a Disposition

Caring about something might involve a significant emotional and motivational commitment, as is often the case with parents who care about their children. But we can also care about someone or something in a much less engaged way: perhaps we care about being prepared for a meeting, but just not very much. In general, in the sense we are after here,

> CARING AS DISPOSITION: *To care about X* is to be disposed to react and act in ways conducive to its going well with X in some regard: disposed to pay attention to information that might be relevant for how well it goes with X and to take the fact that some action would make it go better as reason for that action (i.e. to be motivated to make it go better).

So understood, caring about something is not an activity, but a disposition to engage in various mental and physical activities. Especially when X is a human being, such dispositions are often associated with liking X, or being emotionally committed to the wellbeing of X. However, caring about something, as understood here, does not *necessarily* imply liking it, or having positive emotional reactions to it. One might care about X (a child, the wine, the grading, the climate) not because one likes X, but because one sees it as one's duty to care about X, or stands to gain if it goes well with X (in the relevant regard). Likewise, the hyper-rational and unemotional Mr. Spock (of *Star Trek* fame) might *care more strongly* about a child than someone who has much stronger emotional reactions to the weal and woe of that child, in the sense that he is *more strongly disposed* to take action in her favour. (From this it should be clear that the notion of *caring about* at work here is wider than the notion figuring in care ethics, not being restricted to caring about people to whom one stands in particular social relations, or necessarily related to natural propensities for such interpersonal caring (see e.g. Noddings 1984). Still, we take the resulting general analysis of responsibilities and their relation to requirements to be compatible with the central tenets of care ethics and with the idea that there are requirements to care about others that can normally only be satisfied with the aid of empathy and emotional capacities.)

We will next spell out the idea that responsibilities are requirements to care. Given that caring disposes one to act and react in certain ways, this means that those who discharge their responsibilities to care will also act in certain ways. This, we will suggest, explains how responsibilities give rise to requirements to act.

2.2.2 Responsibilities as Requirements to Care

The first part of our explanation is that responsibilities are *requirements to care*: parents' responsibility for their children is a requirement that they care (in the sense outlined above) about their children. This general idea should be intuitively clear, but it is helpful to say something about the ways in which such requirements can vary, focusing in turn on each of the variables in the following formula:

> RESPONSIBILITIES AS REQUIRED CARING: P's (moral normative) responsibility for an object O (in some regard R of O, within constraints C, to a degree D) consists in the moral requirement that P care, within C, with strength dependent on D, about how well it goes with O with respect to R.

Consider in turn each aspect of this general account of responsibilities:

Bearers of a responsibility (P) are paradigmatically individual moral agents, but responsibilities might also fall on groups of agents or organizations. We say more about this in Sect. 2.2.4.

Responsibilities can have a wide variety of objects (O), such as actions ("making sure there is wine"), activities ("the production of syringes"), domains ("health care"), things ("the car"), states ("John's health"), or individuals ("your son John"). It can even stand for future states of merely possible kinds of individuals, as when one thinks that parents are responsible for the quality of life of potential offspring and that they should therefore choose to create one of those lives rather than another (see Chaps. 7 and 10 of this volume).

Depending largely on the object of responsibility, the caring in question might relate to different ways of going well (different values of Y). When the object is an action, the requirement is typically to care about performing it, and performing it in ways appropriate to the context (*bringing wine, good wine, at the right temperature, at the right time, in suitable containers*, etc.). When the object is a domain, such as a state's health care system, such as the one described by Syurina and Feron in this volume, the requirement is often to care about a variety of interacting values (*accessibility of health care, equality of access, affordability, priority for the seriously ill, focus on preventive health care*, etc.). In the case of most responsibilities, questions about how well it goes with the object of responsibility are multidimensional and complex. Parents will typically have to balance their caring about a child's safety from harm against their caring about her autonomy, and the responsible health care officials will have to balance the accessibility of various kinds of care and economic costs.

Constraints (C) on the required caring are also subject to complex variation. Generally, the relevant caring is constrained by various moral concerns: it is not a sign of failure to care if parents refrain from helping their own child by stealing medication from another, for example. Forms of caring might also be restricted to wider or more narrow contexts. For example, the paediatrician's responsibility for certain children is typically understood as a requirement to care about these children's health in connection with specific doctor's appointments and follow-ups, whereas parental caring is expected to be expressed in a much wider range of contexts.

Finally, the degree (D) of the responsibility and corresponding strength of care required can vary tremendously. If someone is responsible for his child's health as well as for returning a book to the library, we think that the former responsibility carries more weight: if he has to choose between failing to protect the child's health and failing to return the book, the disposition to protect the child's health should be stronger.[7]

Two things are worth mentioning about the relation between RESPONSIBILITY AS REQUIRED CARE and ordinary attributions of responsibility. First, in ordinary parlance the relevant constraints and degrees of responsibility are rarely made explicit, though aspects often are, as when we say that parents are responsible for the child's *safety* or *behaviour*, or that paediatricians are responsible for the *health* of children in their care. Second, what we ordinarily think of as "responsibilities" are relatively significant requirements to care.[8]

2.2.3 *Responsibilities and Requirements to Act*

To understand how responsibilities give rise to requirements of various kinds, we need not only an account of responsibilities in terms of requirements to care, but also an account of the relation between requirements to care and other requirements, in particular requirements to act. Here we propose that in contexts where the required caring about something would ensure (in a normal fashion) that one acts in certain ways—that one feeds a child, or that one does not give it alcoholic beverages—this gives rise to requirements to act in these ways.

Our account builds on four ideas. The first is that under the right circumstances, that we care about something ensures that we perform certain actions, come to know

[7] In addition to different strengths of required care, one might want to distinguish differently strong requirements to care about something to a certain degree, perhaps with reference to the degree of blame that one would be liable to for failing to care about it to that degree.

[8] Hart (1968: 213) tentatively suggests that some duties are "singled out as responsibilities" because they "require care and attention over a protracted period of time". However, it is natural to talk about responsibilities even in the case of very simple but highly significant actions.

certain things, develop certain capacities, have certain emotional reactions, and care about certain other things. This is no accident, of course, given that caring about something is a disposition to notice and act on information in ways conducive to the object of care. For example, the normal way in which caring ensures that an action is performed is by (a) ensuring recognition of reasons suggesting that the action would be conducive to the object of care in the relevant way, where (b) this recognition in turn ensures the action in question. Under the right circumstances, parents' caring about a child ensures that they recognize when it needs nourishing, health care, or emotional support, and that they take action to satisfy these needs. In the same way, caring about something can ensure that we learn facts relevant to the object of care, by ensuring that we recognize or actively seek out the relevant information. Caring about something might also ensure that we have certain emotional reactions and enter into moods of the sort preparing us for the relevant sorts of action, such as fear, anger, determination, curiosity, or compassion. In the case of long-term responsibilities, caring might also ensure that we develop certain capacities that might make caring more efficient, through a general concern for how our behaviour affects the object of care. Finally, caring about something can non-accidentally ensure that we care about any number of other things because we see these as positively affecting the object of care: a parent might care about the temperature in the room or the quality of the food because he cares about his children's health, for example.

Second, in contexts where appropriate caring would have ensured an action or some capacity or emotional or cognitive state in these ways, the lack of such action, knowledge, or capacity is straightforwardly explained by lack of appropriate care. If caring as required would ensure feeding one's child, failure to feed the child must be due to failure to care as required. Conversely, if caring as required would *not* ensure feeding one's child—perhaps it doesn't need feeding at the moment, or there are more important things to attend to—failure to feed one's child might not be explained by failure to care.

Third, it is exactly when one's failure to φ (to act, know, feel, or be capable of something) is the normal, straightforward upshot of lack of appropriate caring that one is liable to moral blame for failing to φ.[9] If one failed to φ for some other reason, this cannot be ground for moral blame.

Fourth, because one is morally required to φ just in case one would be to blame for not φ-ing—this is how we defined moral requirements—there is a requirement that one acts in ways that are ensured in a normal fashion by the required caring.

The upshot of these four ideas is:

[9] For the connection between blame and quality of will, see e.g. Strawson 1982[1962]: Arpaly 2006; McKenna 2012. For the connection between blame and *normal explanations* of objects of blame in terms of agents' quality of will, see Björnsson 2011; Björnsson and Persson 2012, 2013.

REQUIRED Φ-ING AS ENSURED BY REQUIRED CARING: P is morally required to φ if, and only if, were P to care about things in ways that are morally required of P, this would ensure, in normal fashion, that P φ.[10]

REQUIRED Φ-ING AS ENSURED BY REQUIRED CARING captures the close connections between requirements to act, quality of will (whether one cares as required), and blame. The proposal also accommodates the idea that agents are required to φ only insofar as they *can* φ, in a sense that implies strong capacity to φ: proper caring must *ensure* the fulfilment (in a normal way). If morally required caring would only ensure that P *tried* to φ, P would only be required to try. At the same time it allows that one can act as morally required—feed a child, say—without caring as required, or without thinking that the act is required (cf. Ross 1930/2002: 5–6, 42–46).

Given RESPONSIBILITY AS REQUIRED CARING and REQUIRED Φ-ING AS ENSURED BY REQUIRED CARING, it is easy to see how responsibilities can give rise to a variety of requirements to act, depending on the exact nature of the responsibility and the context. But it might also be that very few specific actions are ensured by required caring, generating few requirements to act and leaving great room for the discretion of the responsible party. This will depend on the situation and the nature of the required caring.

2.2.4 Subjects of Responsibility: Individuals, Organizations, Groups

Normative responsibilities are not only attributed to human agents, but also to so-called *corporate agents*, such as corporations, nations, Non-Governmental Organisations (NGOs), and departments of larger organizations, as well as to *groups* of people, such as the parents of a child, the affluent countries of the world, or the bystanders jointly capable of preventing a crime. Such attributions make sense given the nature of normative responsibility and the nature of normative requirements as outlined in the previous sub-sections. Individual agents, organizations and groups are commonly understood as liable to blame for failing to take relevant considerations into account. Corporations are frequently blamed for not taking into account environmental effects of their actions, affluent countries blamed for not doing enough to prevent catastrophic climate change, and pairs of parents collectively blamed for not taking proper care of their children.[11] If it makes sense to

[10] As noted above (n. 4), moral requirements are arguably closely related to moral obligations. For motivation of the corresponding account of moral obligations, see Björnsson 2014: 114–116.

[11] The appropriateness of blame directed at corporations or loosely structured groups of people is a controversial matter. For some defenders, see e.g. Held 1970; French 1984; Rovane 1998; Kutz 2000; List and Pettit 2011. For criticism, see e.g. Corlett 2001; Haji 2006; McKenna 2006, Miller 2006. For our defense, see Björnsson 2011, 2014; Björnsson and Hess 2016. For some relations between individual and group responsibilities or requirements, see Björnsson (forthcoming).

direct blame at corporate agents and groups in these ways, it also makes sense to attribute responsibilities to them.

That an organization or group is responsible for something does not mean that each of its members is responsible for the same thing with respect to the same aspect and within the same constraints. Consider a couple, Alex and Billy, who adopts a child. Before adopting, it was clear to both that Alex's ongoing projects would be hard to combine with being fully engaged in raising the child. It was also clear that Billy wanted a child more and was prepared to take on most of the responsibility for the child, but that Alex would be ready to step in if necessary. Under these circumstances, one might think that Alex and Billy are jointly responsible for the child's welfare, but that the care required from Alex is more circumscribed and conditional than that required by Billy.

The example also illustrates a distinction between *primary* and *secondary* responsibilities, the latter being responsibilities that are conditional on the failure of others to live up to their responsibilities. If Billy's capacity to take care of the child would diminish, or if Billy abandons the child, Alex might now have full responsibility for the child. Similarly, if both Alex and Billy would fail, for lack of capacity or lack of caring, responsibility for the child might fall on some government agency or extended family, who have conditional or circumscribed reasons to care.[12]

2.3 Sources of Special Normative Responsibilities

The main purpose of this section is to give an overview of the most common *sources* of special responsibilities, i.e. the sources that differentiate responsibilities between different agents.[13] To identify the sources of responsibility falls short of exhaustively stating the *conditions* under which someone is normatively responsible, or of providing a full-fledged theory of responsibility. Instead, the question about sources concerns what factors give rise to responsibilities if they are not undermined by other factors. To identify sources of responsibility is thus to identify prima facie principles of responsibility.

We will suggest that there might be at least six distinct sources of normative responsibility, namely (1) capacity, (2) retrospective or causal responsibility,

[12] Related phenomena are those of delegated and transferred responsibility, as when parents leave their child in the care of a babysitter for a few hours, delegating some of their responsibilities, or when they place a child for adoption, transferring responsibilities to the new parents. For discussion of ethical issues with abandoning a child, as compared to putting it up for adoption, see Giordano 2007.

[13] In focusing on special responsibilities, we ignore the question of whether there are universal responsibilities, e.g. a consequentialist responsibility to make the world a better place or deontological responsibilities to, respectively, help and not harm, and how those would be grounded. (So-called "imperfect duties" in deontological theory roughly correspond to what we call "responsibilities". Such duties are not requirements to perform specific actions on every occasion of a certain sort, but can naturally be thought of as requirements to care.)

(3) benefits, (4) agreements, (5) just laws and social norms, and (6) social ties. In distinguishing these sources, we are not implying that there cannot be others. Nor are we assuming that the listed sources will be rock bottom in a final theory of responsibility. Perhaps they are best seen as grounded in more fundamental normative considerations: consequentialist, rule-consequentialist, or contractualist (see e.g. Goodin 1988; Hooker 2000). For example, one might think special responsibilities might be grounded in general impartial responsibilities, motivated by the need for a *moral division of labour*, as proposed by Goodin (1988) in the case of responsibilities towards compatriots. The idea is that we all have general moral responsibilities to all persons, e.g. to help children in need. But for these responsibilities to be best fulfilled, they need to be distributed so that it is clear who should do what for whom. Applied to the case of parental responsibilities, the idea could be to assign primary responsibility for children to parents, while others have mere secondary or other residual responsibilities, because this division of responsibilities would have the best consequences overall. In this section, however, our purpose is to canvas the variety of intuitively plausible sources of responsibilities, without attempting to provide a unified justification.

2.3.1 Capacity and Cost

In some cases, we are responsible for bringing about a morally important outcome because we are particularly well placed to do so. For example, if a child is drowning and you are the only person around who can prevent this, it seems that it is your responsibility to save the child, at least absent extremely strong reasons not to. Often, more than one agent or group is capable of achieving the relevant end. In such cases, their relative capacity might determine the extent to which they are responsible for the outcome. ("With greater power comes greater responsibility".) In general terms, something like the following principle seems plausible (leaving open exactly what counts as a morally important outcome):

> CAPACITY: The party (individual or group) that has most capacity to produce a morally important outcome is, ceteris paribus, responsible for doing so. (Cf. Miller 2001: 460.)

As Miller (2001) points out, capacity to produce an outcome can be regarded as a function of effectiveness (what the agent can do, or how likely she is to succeed) and cost (how costly it is for her to exercise her abilities). This suggests that CAPACITY might be divided as follows:

> LEAST COST: The party for which it is least costly to produce a morally important outcome is, ceteris paribus, responsible for doing so.
> EFFECTIVENESS: The party for which success is most likely is, ceteris paribus, responsible for producing a morally important outcome.

LEAST COST might imply that new technologies (like cell phones) increase responsibility for helping by lowering costs (cf. Wenar 2007). EFFECTIVENESS can explain why being the only one (or one of a few) who can satisfy someone's needs might give rise to special responsibilities, e.g. to one's children.

As formulated, these principles are obviously rough first drafts: The existence of the responsibility as well as its weight will presumably depend on how important the outcome is, and how capable the agent is of achieving it. Moreover, much could be said about the ceteris paribus clause. Depending on what other sources of responsibility are at play, CAPACITY might not apply, and high enough costs might void responsibility altogether. If you can only save the drowning child by risking your own life, you might not be responsible for saving the child, even if you are the person who would be most likely to succeed (Wenar 2007: 260).[14] Finally, the principles, as stated here, are most plausibly concerned with *primary* responsibilities. Corresponding principles could be formulated for secondary responsibilities.

2.3.2 Retrospective and Causal Responsibility

Our past actions might also be a source of normative responsibility. For example, retrospective moral responsibility for harmful outcomes can give rise to a normative responsibility to clean up some mess, deal with the harm, or compensate victims. But other kinds of retrospective responsibility might likewise give rise to normative responsibilities. If one is retrospectively responsible for a situation that *risks leading to harm*, one might be normatively responsible for preventing that harm. Similarly, if one is retrospectively responsible for a situation in which something *needs to be done to prevent harm*, one might be normatively responsible for preventing it even when others would definitely prevent the harm and there thus is no real risk. For example, having knowingly conceived a child, parents might be retrospectively responsible for the existence of a vulnerable being who will be harmed unless it gets copious amounts of help. Even though no harm has been done and no risk created because others will provide help if the parents do not, it might seem that the parents are normatively responsible for helping the child because they brought it into existence. Summing all this up in general terms, it seems that:

> RETROSPECTIVE RESPONSIBILITY: The degree to which P is normatively responsible for setting things right or preventing harm is, ceteris paribus, positively correlated with the degree to which P is retrospectively responsible for the situation calling for these actions.

The most discussed special case is one where P is normatively responsible for compensating some identifiable person or persons for harm for which P is retrospectively responsible (cf. Miller 2001: p 458). However, we might also have normative responsibilities in virtue of being retrospectively responsible for harming or

---

[14] One might think that responsibilities based on capacity and costs follow straightforwardly from universal responsibilities to produce morally important outcomes (cf. note 13 above): if one cares about this, one will care particularly about contributing in cases where one can easily and effectively do so.

endangering some valuable material thing, sacred order, or abstractly identified group (e.g. future generations).

P here can be an individual agent, but also a group jointly retrospectively responsible for the situation. The parents might be jointly normatively responsible for their child in virtue of having jointly knowingly and willingly brought it into existence, and the major industrial countries might be jointly normatively responsible for preventing catastrophic climate change in virtue of having knowingly created or exacerbated the threat.[15]

Something like RETROSPECTIVE RESPONSIBILITY is fairly uncontroversial. But it might also be thought that mere causal responsibility ("contribution" or "participation") can be a source of normative responsibility, even in the absence of moral retrospective responsibility. For example, if you have (blamelessly) harmed someone or someone's property or put someone in danger, you might thereby be normatively responsible, to some degree or other, for setting things right or preventing harm:

> CAUSAL RESPONSIBILITY: The degree to which P is normatively responsible for setting things right or preventing harm is, ceteris paribus, positively correlated with the degree to which P is causally responsible for the situation calling for these actions.[16]

The most familiar case, again, would be one where P is responsible for compensating some identifiable person or persons that P has harmed. But if one accepts a principle like this, it might have wider consequences. For example, even a man and a woman who conceive a child in spite of taking all reasonable contraceptive measures (i.e. unknowingly and unwillingly) could have special responsibilities for the child merely because their having sex was crucial for bringing the child into existence.[17]

As in the case of the capacity principles above, these two responsibility principles are best seen as rough drafts. Both are mere ceteris paribus principles, and the degree to which normative responsibility results is presumably dependent on the degree of causal or retrospective responsibility, which in turn depends on "the way and extent to which the agent is related to the production of that situation" (Haydar 2005: 311).

---

[15] Regarding collective cases, see Ashford 2007; Pogge 2007; cf. also the "polluter pays principle" in the context of environmental impacts. For discussion of joint retrospective responsibility, with a note on how it depends on normative requirements, see Björnsson 2011. Interestingly, it has been suggested that causal responsibility might be unnecessary for retrospective responsibility in collective contexts: the latter might only require that one has *participated* (non-causally) in a harmful collective action (Kutz 2000).

[16] Cf. Barry's (2005, 280) contribution principle, which holds "that agents are responsible for addressing acute deprivations when they have contributed, or are contributing, to bringing them about".

[17] Cf. Archard's (2010: 127) suggestion that "those who cause children to exist thereby incur an obligation that they are adequately cared for".

2.3.3 Benefiting

If you have benefited from someone's help, you might be responsible for reciprocating. For example, you might have a filial responsibility to take care of your aging parents because of what they have done for you over the years. Similarly, according to the so-called *Hart-Rawls principle of fairness*, if one has benefitted from the cooperative scheme of some group, one has an obligation to reciprocate.[18] Benefiting from the actions of one or more others might thus be a source of responsibility:

> BENEFITING FROM HELP: If P has benefited from Q's actions (where Q can be an individual agent, a group, or perhaps a society as a whole), P is, ceteris paribus, responsible for reciprocating, i.e. for helping or supporting Q.

More controversially, one might think that those who benefit from some harm or injustice (e.g. from an unjust system) might have a normative responsibility to help setting things right. For example, the Beneficiary Pays Principle is invoked in the context of climate justice, where it is sometimes argued that "the beneficiaries of activities that cause climate change should shoulder the [relevant] costs" (Page 2012: 304). The costs here might be costs involved in compensating for harm or restoration for injustice, but also costs for preventing harm that would result from the activities in question:

> BENEFITING FROM HARM, INJUSTICE OR DANGER: If P has benefited from something that has brought morally important downsides (injustice, harm, danger) that calls for restoration, compensation or prevention, then P is, ceteris paribus, responsible for contributing to such restoration, compensation or prevention, to a degree dependent on the benefits derived.

In the most typical case, P has benefited from some injustice or harm that some other agent has perpetrated against an identifiable person (or persons) Q and has a resulting responsibility for compensating Q.

Again, the responsibility arising from benefits will depend on a variety of further considerations. Some people defend a restricted version of BENEFITING FROM HARM, INJUSTICE OR DANGER. For example, Caney (2010) suggests that benefiting and causal responsibility only give rise to normative responsibility in conjunction with each other (210; cf. Miller 2001).

2.3.4 Promises, Contracts, and Agreements

Promises, agreements and contracts can also be sources of responsibility. For example, we might be responsible for making sure there is enough wine at a dinner party because we have voluntarily agreed to bring the wine. The same thing holds when the promise or agreement takes the form of a legally binding contract, e.g. in

[18] See Arneson 2013 for discussion and amendments or clarifications of the principle in response to criticism.

business contexts. If you have signed a contract which says that you will deliver a certain service at a certain time, you are not just legally responsible for doing so, but also, in most cases, morally normatively responsible. What these phenomena have in common seems to be that they involve the voluntary agreement or commitment to do something. This suggest that:

> AGREEMENT: If P has voluntarily agreed to φ, then P is, ceteris paribus, responsible for φ-ing.[19]

If one accepts this proposal and takes promises and contracts to be sources of responsibility because they are voluntary agreements, one might also further recognize *implicit* promises, contracts or agreements as sources of responsibility. Whether one should understand "voluntarily agreed" in the context of this principle as including implicit agreements will presumably depend on how one understands explicit agreements and their relation to responsibilities. Perhaps explicit agreements bind by giving rise to justifiable expectations grounded in mutual benefit; perhaps benefiting from some social order (laws, customs) thus constitutes relevant tacit consent and grounds responsibilities, as benefiting without consent would be unfair. The point here, however, is merely to note implicit agreements as a possible source of responsibility.[20]

2.3.5 Laws and Norms

It is easy to imagine different ways of assigning parental responsibilities, and perhaps different ways that could be justly enforced (by law, or by milder forms of social pressure) in a society. Suppose that in one society, there is a norm, justly socially enforced, that genetic parents materially support their offspring (unless incapable, and unless the job has been justly transferred to some other party). Then it might seem that genetic parents in that society would (normally) be responsible for supporting their offspring, in virtue of the existence of that norm. Generalizing, one might think that:

> JUST SOCIAL NORMS: If P is required by just social norms to φ, then P is, ceteris paribus, responsible for φ-ing.

As in the case of other sources of responsibility, the weight of the responsibility might vary (depending on how just the norm is) and this particular consideration might be undercut or outweighed by a variety of others.

While it seems clear that we typically have responsibilities to conform to justly enforced social norms, it is not entirely clear that such norms provide a further

[19] For a helpful introductory discussion, see Jeske 2014.

[20] The problem of tacit consent has received much attention in political philosophy from Hobbes, Locke, and onwards, in connection with the problem of political obligation, i.e. of when a person has a moral duty to obey the laws of her country or state. For an introductory overview of this problem, see Wolff 2006.

source of responsibility, independent of the sources discussed in the previous sections. For example, when socially enforced norms of both formal and informal kinds are voluntarily consented to, explicitly or implicitly, AGREEMENT implies that it is the responsibility of the consenting parties to conform to such norms. By BENEFITING FROM HELP, it is the responsibility of those who have benefited from a cooperative scheme guided by the norms to reciprocate by conforming. And by CAPACITY, it is the responsibility of those who can conform to just norms to do so, insofar as conformity to justly enforced norms contributes to morally important outcomes. Given how difficult it is to find applications of JUST SOCIAL NORMS where neither of these other considerations is at work, it is difficult to find direct intuitive support for thinking that just social norms provide a further independent source of responsibilities. For this reason, we refrain here from taking a stand on whether it does.

2.3.6 Roles and Special Relationships

Responsibilities are often associated with roles. In fact, the association is so strong that what we call normative responsibility is sometimes referred to as "role-responsibility". Hart (1968) introduced this notion by pointing out that

> whenever a person occupies a distinctive place or office in a social organization, to which specific duties are attached to provide for the welfare of others or to advance in some specific way the aims or purposes of the organization, he is properly said to be responsible for the performance of these duties, or for doing what is necessary to fulfil them. Such duties are a person's [role] responsibilities. (212)[21]

This makes it natural to think of roles as independent sources of normative responsibility, i.e. to assume that we have a number of special responsibilities *in virtue of* our social or professional roles.

Whether this is plausible depends on how roles are understood. Suppose that we understand roles as themselves constituted by moral requirements and responsibilities. Then the role would not be so much the source of these requirements as their upshot. But suppose, instead, that roles are constituted by laws or other social norms tied to positions in a social system. Then roles could indeed be sources of requirements, either because the norms in questions were just (by JUST SOCIAL NORMS), or because people agree with the norms constituting the roles when they occupy these roles (by AGREEMENT).[22] This seems particularly clear in the case of professional roles that are largely defined by contracts and laws, and perhaps many other roles, including those of a friend, partner, or parent, can at least partly be similarly defined by social norms. Moreover, norms associated with particular roles might give those

[21] Here, Hart extends the sociological notion of a role "to include a task assigned to any person by agreement or otherwise" (p. 213). However, Hart takes only a subset of the duties connected to roles to constitute responsibilities (cf. n. 9).

[22] See e.g. Hardimon 1994.

occupying the roles certain social capacities, capacities that are in turn sources of responsibility (by CAPACITY). Relatedly, roles might typically be associated with and perhaps even defined by various non-normative capacities to care. Parents will typically be particularly strongly disposed to care about their children, and friends to care about their friends, and weaker social relations can give rise to relations of solidarity among members of social groups, making it easier for them to help.[23] Correspondingly, children have emotional needs in relation to their regular caretakers that might not be easily satisfied by others. The point here is not the trivial one that people in close relationships do in fact care about each other. Rather, the idea is that such relationships might give rise to *normative* responsibilities, e.g. through CAPACITY. In giving rise to responsibilities for these sorts of reasons, however, roles are clearly parasitic on the other sources already mentioned. It is questionable whether they provide a further, independent sources of responsibilities.

Similar things can be said about the suggestion that community membership or special social relations of various kinds are sources of responsibilities. For example, David Miller (2001: 462) suggests that remedial responsibilities—responsibilities to put bad situations right—are in part distributed based on "special ties of various kinds such as those that exist within families, collegial groups of various kinds, nations, and so forth". These ties, he thinks, can arise from "shared activities and commitments, common identities, common histories, or other such sources". Depending on how they are understood, these characterizations raise the same question as appeals to roles. If the ties are understood in normative terms, as requirements, the question is from where they arise. But it is also easy to see how social relations can give rise to requirements, given the sources we have already covered. Many social or intimate relationships are subject to *just social norms*, or expectations involving voluntary *agreement*, or relations involving caring and corresponding emotional needs yielding special *capacity* to help, or relations of *benefit*. If these are the grounds of responsibilities, however, social ties do not provide sources beyond what has already been covered.[24]

Still, we want to suggest that some social ties are further sources of responsibilities. Consider a case involving a minimal history of cooperation:

> *No complete stranger*: Two people, Q and R, have fallen into a river, and you can reach out and help one onto dry ground. Both are at equal distance and would be equally easy to help, but whereas Q is a complete stranger, you recognize R from yesterday. Someone had asked how to get to City Hall, and you and R had both provided partial answers in a brief but cooperative effort.

In this situation, it seems to us that your responsibility to help R is stronger than your responsibility to help Q: the mere fact that you share a brief cooperative history makes a difference. Still, it is not clear that any of the other sources mentioned here can explain this. Moreover, the extent of the cooperative history seems to matter: if you have had some cooperative interaction with Q but a much more extensive

[23] Cf. Feinberg (1968), who takes relations of solidarity to ground collective responsibility.

[24] Similar things can be said about relations figuring in Iris Marion Young's (2013) "connection model" of forward-looking responsibility for injustice.

interaction with R, this too would make a difference, and one that could trump other sources of responsibilities. Perhaps, then, at least some kinds of social ties constitute further sources of at least remedial responsibilities:

> SOCIAL TIES: If Q and R need help and P is more strongly linked to Q than to R by relevant social ties (such as histories of cooperative interaction), then P has greater (remedial) responsibilities towards Q than towards R, ceteris paribus.

For reasons of space, we refrain here from trying to determine exactly how to understand the relevant social ties, or whether they extend beyond cooperative histories to other social or communal relations of the sort mentioned by Miller.[25] The formulation of SOCIAL TIES, perhaps even more than those of the other principles discussed in this section, is best understood as a first draft. Much more can be said about the nature of the relevant social ties and their relation to roles, community membership and intimate relations, about the weight and contents of the resulting responsibilities, and about further extensions or qualifications of the principle.[26] Still, even restricted to this one kind of social tie, the principle might have important consequences for responsibilities towards friends, lovers, spouses, and coworkers, as well as towards one's parents or children: all these relations typically involve rich histories of cooperative interaction.

2.4 Conclusion

In Sect. 2.2, we proposed that moral normative responsibilities should be understood as moral requirements to care about the object of responsibility. It is such requirements, we suggested, that give rise to the variety of other requirements that seem to follow from our responsibilities, including requirements to act, and requirements to care about acting in certain ways. In Sect. 2.3, we have briefly discussed what we take to be the more obvious kinds of sources of special responsibilities.

As we have indicated, most of these kinds of sources seem relevant for various forms of parental responsibility, and for the theme of this volume.[27] But the sources also interact with each other and raise numerous complications, many of which are discussed in other chapters. How much weight does one type of responsibility have compared to other responsibilities? What exactly are parents responsible for, and what are the constraints (see Chaps. 3, 5, 6, 7, 9 and 12, in this volume)? How much information can, or need they seek; how much can or should they control (see Chaps. 4, 9 and 12, in this volume)? What responsibilities do parents have towards

---

[25] Cf. Bell (2013), section 3, for a useful distinction between three types of communities, namely communities of place, communities of memory, and "psychological communities", or "communities of face-to-face personal interaction governed by sentiments of trust, co-operation, and altruism".

[26] Cf. Scheffler (1997), who suggests that special relationships only give rise to responsibilities when people have reason to value these relationships.

[27] For an earlier discussion of some of these issues, see Brennan and Noggle (2007).

a fetus, or to children at various stages of development (see Chaps. 9, 10, 11, 13 and 14)? When parents can control which child comes into existence, are they responsible for choosing children that can have a higher quality of life, or are they merely responsible for the quality of life of the child they actually bring into existence (see Chaps. 10 and 11)? What are the rights and responsibilities of others when parents fail to discharge their own responsibilities (see Chaps. 4 and 8)? Though we have done nothing in this chapter to answer these questions specifically, what we have provided are some general perspectives that can hopefully guide the reader in thinking further about these issues.

Contributorship Statement Björnsson proposed the overall structure of the paper and is primarily responsible for Sect. 2.2, on the structure of normative responsibilities. Brülde is primarily responsible for Sect. 2.3, on the sources of normative responsibilities. Both authors have contributed to the chapter as a whole and approved the final version.

References

Archard, D. 2010. The obligations and responsibilities of parenthood. In *Procreation and parenthood: The ethics of bearing and rearing children*, ed. D. Archard and D. Benatar, 103–127. Oxford: Oxford University Press.

Arneson, R. 2013. Paternalism and the principle of fairness. In *Paternalism*, ed. C. Coons and M. Weber, 134–156. Cambridge: Cambridge University Press.

Arpaly, N. 2006. *Merit, meaning, and human bondage: An essay on free will*. Princeton: Princeton University Press.

Ashford, E. 2007. The duties imposed by the human right to basic necessities. In *Freedom from poverty as a human right: Who owes what to the very poor?* ed. T. Pogge, 183–218. Oxford: Oxford University Press.

Barry, C. 2005. Applying the contribution principle. In *Global institutions and responsibilities: Achieving global justice*, ed. C. Barry and T. Pogge, 280–297. Oxford: Blackwell.

Bell, D. 2013. Communitarianism. In *The Stanford encyclopedia of philosophy* (Fall 2013 Edition), ed. Edward N. Zalta. URL = http://plato.stanford.edu/archives/fall2013/entries/communitarianism/.

Björnsson, G. 2011. Joint responsibility without individual control: Applying the explanation hypothesis. In *Compatibilist responsibility: Beyond free will and determinism*, ed. Jeroen van den Hoven, I. van de Poel, and N. Vincent, 181–199. Dordrecht: Springer.

Björnsson, G. 2014. Essentially shared obligations. *Midwest Studies in Philosophy* 38(1): 103–120.

Björnsson, G. Forthcoming. Individual and shared obligations. In *Philosophy and climate change*, ed. M. Budolfson, T. McPherson, and D. Plunket. Oxford: Oxford University Press.

Björnsson, G., and K.M. Hess. 2016. Corporate crocodile tears? On the reactive attitudes of corporations. *Philosophy and Phenomenological Research*, Early View. doi: 10.1111/phpr.12260.

Björnsson, G., and K. Persson. 2012. The explanatory component of moral responsibility. *Noûs* 46(2): 326–354.

Björnsson, G., and K. Persson. 2013. A unified empirical account of responsibility judgments. *Philosophy and Phenomenological Research* 87(3): 611–639.

Brennan, S., and R. Noggle (eds.). 2007. *Taking responsibility for children*. Waterloo: Wilfrid Laurier University Press.

Caney, S. 2010. Climate change and the duties of the advantaged. *Critical Review of International Social and Political Philosophy* 13(1): 203–228.

Corlett, J.A. 2001. Collective moral responsibility. *Journal of Social Philosophy* 32(4): 573–584.

Feinberg, J. 1968. Collective responsibility. *The Journal of Philosophy* 65(21): 674–688.

French, P.A. 1984. *Collective and corporate responsibility*. New York: Columbia University Press.

Giordano, S. 2007. Crimes and misdemeanours: The case of child abandonment. *Journal of Medical Ethics* 33(1): 28–34.

Goodin, R.E. 1988. What is so special about our fellow countrymen? *Ethics* 98(4): 663–686.

Goodin, R.E. 1995. *Utilitarianism as a public philosophy*. Cambridge: Cambridge University Press.

Haji, I. 2006. On the ultimate responsibility of collectives. *Midwest Studies in Philosophy* 30(1): 292–308.

Hardimon, M. 1994. Role obligations. *The Journal of Philosophy* 91(7): 333–363.

Hart, H.L.A. 1968. *Punishment and responsibility*. Oxford: Oxford University Press.

Haydar, B. 2005. Extreme poverty and global responsibility. In *Global institutions and responsibilities: Achieving global justice*, ed. C. Barry and T. Pogge, 310–322. Oxford: Blackwell.

Held, V. 1970. Can a random collection of individuals be morally responsible? *The Journal of Philosophy* 67(14): 471–481.

Hooker, B. 2000. *Ideal code, real world: A rule-consequentialist theory of morality*. Oxford: Clarendon Press.

Jeske, D. 2014. Special obligations. In *The Stanford encyclopedia of philosophy* (Spring 2014 Edition), ed. Edward N. Zalta. URL = http://plato.stanford.edu/archives/spr2014/entries/special-obligations/.

Kutz, C. 2000. *Complicity: Ethics and law for a collective age*. Cambridge: Cambridge University Press.

List, C., and P. Pettit. 2011. *Group agency: The possibility, design, and status of corporate agents*. Oxford: Oxford University Press.

McKenna, M. 2006. Collective responsibility and an agent meaning theory. *Midwest Studies in Philosophy* 30(1): 16–34.

McKenna, M. 2012. *Conversation and responsibility*. New York: Oxford University Press.

Miller, D. 2001. Distributing responsibilities. *Journal of Political Philosophy* 9(4): 453–471.

Miller, S. 2006. Collective moral responsibility: An individualist account. *Midwest Studies in Philosophy* 30(1): 176–193.

Munthe, C. 2011. *The price of precaution and the ethics of risk*. Dordrecht: Springer.

Narveson, J. 2007. Parental responsibility. In *Taking responsibility for children*, ed. S. Brennan and R. Noggle, 19–45. Waterloo: Wilfrid Laurier University Press.

Noddings, Nel. 1984. *Caring: A feminine approach to ethics and moral education*. Berkeley: University of California Press.

Page, E.A. 2012. Give it up for climate change: A defence of the beneficiary pays principle. *International Theory* 4(2): 300–330.

Pogge, T. 2007. Severe poverty as a human rights violation. In *Freedom from poverty as a human right: Who owes what to the very poor?* ed. T. Pogge, 11–53. Oxford: Oxford University Press.

Ross, W.D. 1930/2002. *The right and the good*. Oxford: Clarendon Press.

Rovane, C.A. 1998. *The bounds of agency: An essay in revisionary metaphysics*. Princeton: Princeton University Press.

Scheffler, S. 1997. Relationships and responsibilities. *Philosophy and Public Affairs* 26(3): 189–209.

Schmidtz, David, and Robert E. Goodin. 1998. *Social welfare and individual responsibility*. Cambridge: Cambridge University Press.

Smith, A.M. 2012. Attributability, answerability, and accountability: In defense of a unified account. *Ethics* 122(3): 575–589.

Strawson, P.F. 1982[1962]. Freedom and resentment. In *Free will*, ed. G. Watson, 59–80. Oxford: Oxford University Press.

Vincent, Nicole. 2011. A structured taxonomy of responsibility concepts. In *Moral responsibility*, ed. N.A. Vincent, I. van de Poel, and J. van den Hoven, 15–35. Dordrecht: Springer.
Wenar, L. 2007. Responsibility and severe poverty. In *Freedom from poverty as a human right: Who owes what to the very poor?* ed. T. Pogge, 255–274. Oxford: Oxford University Press.
Wolff, J. 2006. *An introduction to political philosophy*, revised edition. Oxford: Oxford University Press.
Young, I.M. 2013. *Responsibility for justice*. Oxford: Oxford University Press.

Chapter 3
Disorders Are Reduced Normativity Emerging from the Relationship Between Organisms and Their Environment

Anna M.T. Bosman

Abstract The rise of modern medicine has led to a distinction between physical illness and health based on physiological measures. Psychiatry, the study of mental disorders, contingent on the medical model, attempts to establish a science that distinguishes between normal and pathological conditions. Following the work of the medical doctor George Canguilhem, this paper provides a conceptual analysis of disorders. Along the lines of Canguilhem, I will argue that medicine and psychiatry cannot be sciences in the sense in which physics or chemistry are sciences (that is, with a claim of objectivity or being value-free), because to establish that someone is healthy or has a pathology requires a normative act – and thus a departure from any ideals of objectivity. Health means the ability to maintain life given the existing circumstances, whereas pathology is the diminished possibility of adaptation to the environment. Because health and pathology are the irreducible result of the relationship between the organism and its environment, an individual cannot objectively be assigned a pathology or disorder.

Keywords Disorder • Pathology • Normativity • Organism-environment relationship • Canguilhem

---

A.M.T. Bosman (✉)
Faculty of Social Sciences, Behavioural Science Institute, Radboud University Nijmegen, Nijmegen, The Netherlands
e-mail: a.bosman@pwo.ru.nl

K. Hens et al. (eds.), *Parental Responsibility in the Context of Neuroscience and Genetics*, International Library of Ethics, Law, and the New Medicine 69,
DOI 10.1007/978-3-319-42834-5_3

3.1 Introduction[1]

The development and release of the most recent Diagnostic and Statistical Manual of Mental Disorders (DSM-5) of the American Psychiatric Association (APA 2013) has spawned a worldwide and heated debate on the meaning and value of psychiatric diagnoses. This debate covers two main categories, a content (BPS 2011; Frances 2012; Hacking 2013) and a conceptual discussion (e.g., Greenfield 2013a; Grohol 2013; van Os 2014).

With respect to *content*, detailed discussions took place on the issue of whether or not certain diagnoses were still considered eligible as separate categories. For example, Asperger and PDD-NOS were separate classifications in the DSM-IV, they are no longer in the DSM-5 and are now categorised as 'Autistic Spectrum Disorder' (Frances 2012). At the same time, interest groups lobbied for the recognition of a specific problem as an official disorder. The Sensory Processing Disorder Foundation (2012) unsuccessfully campaigned for the entry of Sensory Processing Disorder in the DSM-5. Critique also targeted the increasing range of behaviours considered to be *abnormal*. For example, normal grief is now called Major Depressive Disorder and childhood tantrums are now readily classified as Disruptive Mood Dysregulation Disorder (Frances 2012).

Conceptual disputes concern a rather varied range of topics fuelling the controversy during these historical times in psychiatry, as Decker (2010) characterised its protracted development. The ontological status of mental disorders, that is, whether or not their ontological status is similar to that of medical diseases, is one of the most important of these topics. The causes of a large number of medical diseases, certainly not all, are known, whereas those of mental diseases are not. Although some psychiatrists maintain that mental disorders are in fact brain diseases, van Os (2014) argues that we are far away from a clear connection between brain abnormalities and overt, pathological, behaviour. Another problem that troubles the discussion is the absence of a unified theory of mental disorders (Frances and Widiger 2012). A final dispute was put forward by Greenfield (2013b) who states that the fundamental role of culture in the aetiology of mental problems has been severely neglected. Culture and environment can have a significant role in the causation and development of disorders.

Unlike Greenfield, I intend to show that disorders are the result of inherently normative acts that emerge from the irreducible relationship between the organism and the biological environment. This view on disorders has clear implications for the debate on responsibilities for parents and caretakers of children who have problems coping with life. Rather than dealing with whether or not X is a disorder, I will

---

[1] This chapter is loosely based on a paper presented on the 9th of December 2013, at the symposium "Parental Responsibility in the Context of Neuroscience and Genetics", organized by the Department of Health, Ethics, and Society of Maastricht University in Maastricht, the Netherlands. I would like to thank Ton Mekking for his contribution on the issue of measuring temperature, Marianne Reuling for an in-depth discussion of the societal implications and Nora Loretan for her careful reading of the manuscript.

make a conceptual analysis of the terms 'normal' and 'pathological' or 'normal' and 'abnormal'. What does it mean when we say that behaviour is pathological or abnormal? The French medical doctor and philosopher George Canguilhem (1904–1995) wrote a profound and radical book about this issue.[2] My contribution is much indebted to his thinking.

3.2 How Health Became the Norm

The modern approach to diseases originates from Hippocrates, the father of Western medicine, who maintained that diseases occur naturally rather than as the result of an act of God. He stated that nature is in harmony or equilibrium and disease is the disturbance of this harmony. Because equilibrium in the organism consists of the four humours: blood, phlegm, yellow and black bile, it is the disturbance of these humours that results in disease. Disease is not so much the disequilibrium itself, but the effort to restore equilibrium. The organism develops a disease to re-establish the proper distribution of humours; it is a response to an internal battle between forces. The classical Greek view on health reveals that sickness and health are two distinct qualities, caused by a shortage or excess of something, namely, humour, and they refer to a different organisation of the organism (Canguilhem 1966). Note that this view reveals that disease is not disorder; it is another kind of order.

The nineteenth century saw a change. The discovery of germs led to the view that a disease *enters* the person or organism. Something pathogenic that comes from the outside causes *malheur*. The fact that we can detect a germ (i.e., bacteria or virus) in the body appears to guarantee an ontological cause of the disease. According to this view, disease is the result of the battle of the organism against an enemy from the outside.

Modern medicine evidences the existence of an illness from laboratory analyses that reveal either an elevated or reduced level of something in the blood, often caused by an intrusion from the outside. Thus, illnesses or diseases were and still are coined in terms of either a lack or an excess of something in the patient, and pathology became a deviation from the standard. For example, a person suffering from diabetes has elevated levels of blood sugar. Even if the person has all the symptoms of someone suffering from diabetes, without an abnormal level in the blood the patient will not be diagnosed with diabetes.

To establish whether, for example, blood-sugar levels are too high or too low, norms were needed. Physiology, the new science that arose in the nineteenth century,

---

[2] I will be referring to two books. The first was published integrally in 1966 in French and translated into English in 1978. Part one was originally published in 1943 and constitutes his doctoral thesis, 'Essay on some problems concerning the normal and the pathological'. Part two was written between 1963 and 1966 and is called 'New reflections on the normal and the pathological'. When I refer to this book, I will use 1966 as the date of publication. Referring to page numbers, I will add 1989, the printed copy of the book I used, published by Zone Books. The second book contains selected writings from Canguilhem edited by François Delaporte, published in English in 1994.

aimed at precisely that: establishing norms for the 'normal' function of living things. The way to go about this had been paved by the work of Adolphe Quetelet (1796–1874), the father of biometry who used the *Gaussian* or *Normal* distribution to justify the use of statistical averages and the deviation from it. The statistical mean, obtained from empirical research, is equated with the norm and the more a characteristic deviates from this statistical norm, the rarer it becomes. With respect to the goal of my paper it is important to emphasize that Quetelet explicitly claims that the statistical mean has ontological status, the statistical mean does represent something real in the world (see Canguilhem 1966/1989: 158).

Before discussing the problems with the definition of pathology or disorder as an attribute whose value deviates sufficiently from a norm, we first need to establish what allows for a proper or valid measurement. This will appear to be particularly important with respect to psychiatric of psychological concepts.

3.3 Valid Measurements

3.3.1 Measuring Physical Phenomena

To assess some attribute, we need a valid tool or test. The example of the thermometer will serve as an illustration to define the characteristics of a valid measurement. A thermometer is a tool that measures the temperature of a body (e.g., the environment, a gas, the bathwater, an organism, etc.). To measure the heat or hotness of a body, physicists rely on a theory with respect to heat and energy. The assumption in the development of thermal physics is that bodies consist of molecules and these molecules move or vibrate. Three important implications are drawn from this: (1) the speed with which the molecules move can vary, (2) the movement of the molecules contains energy, called kinetic or thermal energy, and (3) bodies with higher kinetic energy (the hot body) can release their energy to bodies with a lower level of kinetic energy (the cold body). Figure 3.1 represents an example of a hot body (black molecules) next to a colder one (striped molecules). One additional fact is required to understand the workings of a thermometer, namely, bodies, particularly visible in liquids, expand when kinetic energy is added to the body.[3]

Assume that the black molecules represent the temperature of the body to be measured. The striped molecules represent the thermometer (also a body). When the thermometer is placed next to the body of which the temperature has to be assessed, the body with the highest kinetic energy (the hotter one) will release its energy to the one with the lower level of kinetic energy (the colder one), until they both have the same level of kinetic energy. When the kinetic energy of a body increases – in the case of a thermometer the body is usually a liquid (mercury or

[3] This knowledge is expressed in the first law of thermodynamics: $P*V = n*R*T$. The pressure (P) of a body multiplied by its volume (V) is proportional to the amount of gas (n) multiplied by the gas constant R and its temperature T.

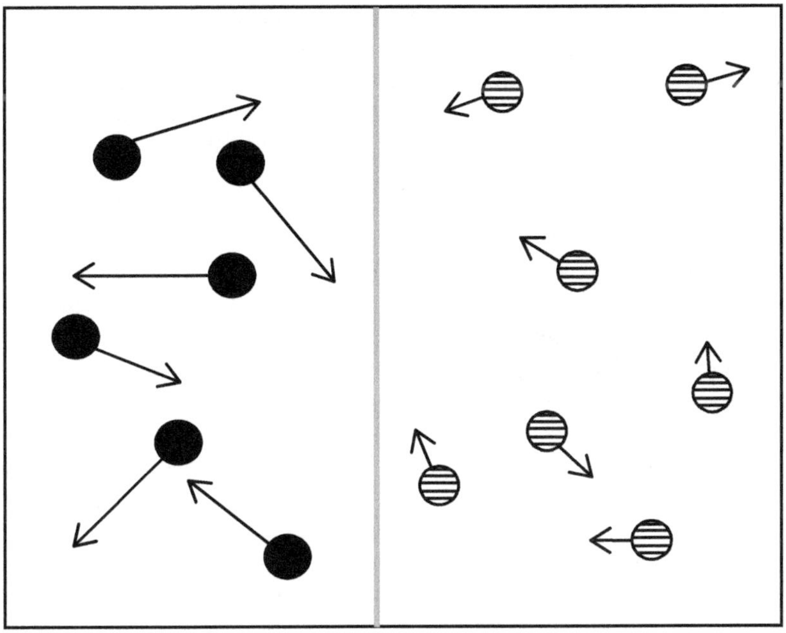

Fig. 3.1 Vibrating molecules in which the *right panel* is the thermometer and the *left one* the body to be measured

alcohol) – the liquid will expand and take up more volume. This is visible in the rise of the liquid in the tube. The more the liquid in the thermometer rises the more kinetic energy is transferred from the measured body to the thermometer. When the liquid stops rising (or falling for that matter), it means that the average kinetic or thermal energy of the two bodies is identical and we have an indication of the temperature of that body.

Of course, we need a calibration procedure to be able to obtain norms for the amount of kinetic energy (i.e., the temperature) of measured bodies. A well-known thermometer is the one developed by the Swedish astronomer Anders Celsius in 1742. Celsius has set the melting point of ice at 0 degrees and the boiling point of water at 100 degrees. Within this range, the increase in temperature develops more or less linearly. Dividing the length of the tube in 100 equal steps makes it possible to measure temperatures according to shared norms. The fact that more than one type of thermometer was developed does not threaten the validity of our measurement tool (e.g., Celsius, Fahrenheit, Réaumur, Kelvin). After all, scores on each of these measurement devices can be transformed into one another, which evidences that the norms of the different thermometers are related to some fundamental underlying principle.[4]

[4] To compute the reading of a Fahrenheit thermometer into Centigrade we take the measurement of the Fahrenheit thermometer, first withdraw 32 from it, and then multiply this number with 5/9.

At this point, I need to point out an important caveat that I will return to later: the temperature reading of the body is in fact the reading of the kinetic energy of the body combined with that of the thermometer, the measurement device. Before the thermometer was put next to the body, the body did not release energy to the thermometer. To be able to obtain information from the body, it has to do so. This means that if the body had a higher temperature than the thermometer, the reading will be slightly too low. This loss of information will be negligible in case of measuring the temperature of water in a bathtub or that of a human body, but not when we want to measure the temperature of a drop of water with a regular household thermometer. The essential point taken from this is that measuring something is intervening in the process at hand. Thus, a measurement or a score is always the result of the interaction between the entity, phenomenon, or process we are interested in and the measurement device. At macro-scale physical levels, this interaction usually does not create problems, but it does on micro levels, such as in quantum physics.

This example from physics reveals that once we have some kind of recognition of an attribute out there that needs to be measured and a theory of what is actually happening between the attribute and the measurement or the test score (in this case the transfer of kinetic energy and the fact that bodies expand when their kinetic energy increases), we are able to make sense of test scores (i.e., the reading of the thermometer).

Consequently, the construction of a measurement device depends on knowledge of the workings of the phenomenon to be measured. My elaborate discussion of how temperature is measured was inspired by the seminal work of Borsboom et al. (2004) on the concept of validity. Their definition reads:

> A valid test can convey the effect of variation in the attribute one intends to measure. This means that the relation between test scores and attributes is not correlational but causal. A test is valid for measuring an attribute if variation in the attribute causes variation in the test scores (p. 1067).

This is exactly what a thermometer does. There is an attribute, namely kinetic energy. The amount of kinetic energy is causally related to the reading of the thermometer (test scores), because the amount of kinetic energy is proportional to the score on a thermometer. Moreover, Borsboom et al. (2004) use of *validity* coincides with that of most social scientists, namely, a valid test measures what it is supposed to measure. Thus, there exists an attribute in reality (ontological claim), which reveals the effect of variation of the attribute (there is a referent), because the attribute causally affects the outcome of the measurement protocol (hence it is not based on correlation, but on causation).

3.3.2 Measuring Psychological Phenomena

What does this mean if we want to measure disorder or pathos or just some psychological phenomenon? We need to be sure that the attribute we want to measure does in fact exist. We also need a theory to explain the variation in test scores that can be ascribed to variation in the attribute and we need a tool that is capable of picking up on the variation of the attribute to be measured. The ontological claim as well as the corresponding conclusive theory is precisely what is lacking in all tests developed in psychology and psychiatry.

Take for example the widely used intelligence test. Its statistical construction provides the percentile rank, based on standard scores, of the one who took the test. A percentile rank is an indication of the percentage of people of a particular age group who scored lower on that test than this particular individual. For instance, an IQ of 120 means that 91 % of the age group scores lower and an IQ of 65 means that only 1 % scores below that level.

The first question that needs to be answered is whether or not intelligence exists. What is its ontological status? For that we need a theory. Unfortunately, there is no consensus on intelligence and although many test developers maintain they have a theory of intelligence, I dare to say that they are confusing description with theory. The examples that follow are taken from Kaufman's second chapter (2009). Binet and Simon stated that intelligence consists of three different aspects: direction, adaptation, and criticism. Guilford defined intelligence in terms of three dimensions: (1) intellectual[5] processes required to solve a problem, (2) the content of the problem, and (3) the products or how stimuli are organized. The most interesting one is provided by David Wechsler, developer of the 'Wechsler Intelligence Scale for Children' (WISC). His definition of intelligence, as we will see below, closely resembles Canguilhem's concept of health: intelligence, he explains, is "...a person's overall capacity to understand and cope with his or her environment" (Kaufman 2009: 43).

Another indication of the limited theoretical value of the concept of intelligence is the fact that many different intelligence tests exist (WISC, WAIS, Stanford-Binet Intelligence Scale, Woodcock-Johnson III, Raven, K-ABC, etc.) and scores on these tests do not necessarily coincide. Kaufman (2009: 152) reveals differences in scores of up to 22 IQ-points obtained by one participant, which amounts to being evaluated as having above average IQ on one test and below average on the other. Unlike the thermometer, different intelligence tests do not provide identical IQ-scores. The argument that measurement error should be taken into consideration is not an acceptable explanation, because thermometers also produce measurement errors. Based on a sound theory, it can be argued that thermometers as well as sphygmomanometers (devices that measure blood pressure) are causally related to their respective test scores, whereas IQ and reading skill are only correlationally related to a test score.

After this exposé on the difference between attributes that are believed to be real (or at least for the existence of which we find evidence) and attributes that are questionable, another equally important issue requires discussion. Suppose it was possible to measure psychological phenomena validly (after all, there are numerous scientists who claim they can, given the ever-increasing number of 'acknowledged'

[5] Note that Guilford uses the word intellectual as part of the explanation for intelligence. Using the attribute to be explained (explanandum) in the explanans is logically invalid.

psychological tests), then the question arises *why* and based on *what* are we deciding that some value of a measurement is too high or too low. In the end, the question that needs to be answered in the context of this chapter is: How are we able to assess that psychological and for that matter psychiatric phenomena deviate sufficiently from some standard in order to call them a disorders or disabilities? I discuss these issues in the next paragraphs.

3.4 Normal and Abnormal

When measuring temperature, mass, or distance, we attribute some number in pre-defined units (centigrade, kilogram, and meter, respectively) to the body or process of interest. There is nothing in the number on a thermometer, speedometer, or scales that indicates that the score is too high or too low; it just provides a number. It is the human observer who appraises its meaning. Appraisal is always contextually bound, *too high* is always too high with respect to something. A car (body of interest) that drives at a speed of 120 km per hour is too fast on a road with a speed limit of 100 km per hour (given that the driver tries to avoid a fine), and too slow on some stretches of a Formula One race circuit (given that the driver wants to win). In other words the number is always appraised or evaluated with respect to the context and the goal. There are no inherent norms in proper measurement devices.

What about psychological measurements? Let us return to the IQ-test. The score on an IQ-test is not based on some pre-defined unit of measurement. The number that results from administering an IQ-test is based on the results of a large group of people with a certain age who took the test at some point. Thus, the reading of an IQ-test is the result of some-pre-established norm that constitutes the test. The norm is the average score on the test (for which there is no real entity or theory that links the test score with some attribute) and scores below or above average are considered deviations. The IQ-test, in fact all psychological tests, have their norms built-in. The scores only reveal the ordering of people with respect to one another. Note that ascertaining deviations from the norm, which are used to argue that a person with an IQ-score way below average is not eligible for regular education, is the second step in the normative process of evaluating people, not the first.

The fact that all psychological tests are based on statistical values, usually averages, is not necessarily a problem if their purpose is, for example, that we want to find the best person for a task (provided that the test measures what it is supposed to measure). However, most psychological tests are used to assess whether someone deviates too much from the average, and this assessment is used to deem the person abnormal[6] for that particular psychological trait or disposition. Unlike in the case of speed, where the number that results from the test has to be appraised in the context of the situation, the result from a psychological test is the appraisal itself. Therefore,

[6] Note that I use the word abnormal here because of its meaning 'departing from the normal' similar to Canguilhem's (1966) usage of the term.

psychological testing is a normative act without taking context into consideration.[7] All so-called standardized tests come with norms by which to assess the outcome. The Child Behaviour Checklist (CBLC; Achenbach 1991) for the identification of problem behaviour in children determines whether the child's score reveals normal, borderline, or clinical behaviour; the latter being the pathological condition.

3.4.1 *The Etymology of Normal*

Normal in the sense used in psychological tests apparently reveals healthy behaviour. But to what extent is the origin of the word *normal* related to health? A linguistic or etymological analysis reveals something interesting. The noun *norm* comes from the Latin *norma*, referring to 'a carpenter's square, rule, or pattern', which has been around at least since 1670 in the English written language. *Normal*, thus, is the adjective of norm deriving from the Latin normalis 'in conformity with rule' or 'made according to a carpenter's square'. In 1890, normal was first used in text referring to 'usual state or condition' and in 1894 to 'normal person or thing' (Harper's Online Etymology Dictionary).

Note that *norm* and *normal* were originally used to refer to a tool, an instrument that was used by carpenters, the so-called T-square; the reference is an artefact, something rather unusual. After all, one does not find T-square shapes in nature. In that respect one could say that the norm is actually the unusual rather than the usual or common: normal is something that may in fact not exist. Just like, as we all know, the average person does not exist. Of course, a linguistic analysis of the developing concepts of *norm* or *normal* does not necessarily discredit its current-day usage. It is nevertheless revealing that *normal* used for people has something to do with shaping people into some prototype, which the T-square is. The emergence of equating normal with healthy may originate from *normal* meaning 'usual state or condition', assuming that health is the usual condition. The fact that health is a desired condition does not, however, guarantee that it is usual.

Establishing normality or health based on (standard) deviations from a statistical quantity such as the mean or modus does not provide us with an objective measure of the condition of a person. There are many situations in which deviations from the norm are not viewed as unhealthy. Take for instance a high score on an intelligence test. A score of three standard deviations above the mean is considered a rather 'healthy' albeit abnormal intelligence. Only when the IQ-score is substantially below the mean, will it be considered an abnormal, unhealthy score.

Why is a high score on an IQ-test considered healthy, despite it being abnormal, and a low score, abnormal and unhealthy? A high IQ often provides the individual with more and better opportunities in life, whereas a (very) low IQ is often the fore-

[7]The fact that the context, people of a certain age taking the test, was used to establish the norm does not invalidate the argument that the result is interpreted as an 'objective' context-free measurement.

bode of a difficult future usually requiring long-term care. The qualification 'often' reveals that people with high IQ's do not necessarily lead healthy lives and those with low IQ's unhealthy. There are numerous exceptions. In fact, the literature on giftedness demonstrates quite clearly problems that people with high IQ's may encounter (e.g., Seeley 2004). Like in medicine, the statistically abnormal may be healthy, and the statistically normal can be sick (e.g., 70 % of a population may contract the flu). An example is 'Dyschromic spirochetosis'. This condition is characterized by coloured spots on the skin and was so prevalent in the South American Indian tribe Pinta, that those who did not have the spots were considered sick (pathological) and could not marry (Sedgwick 1972). Although a fever is relatively unusual, it is at the same time a healthy response to something that went wrong, because the reaction of the body to increase the temperature enhances the immune system. In sum, to equate health with normal in the statistical sense does not appear to be a fruitful direction, not even in medicine,[8] because deviations from statistical averages may never lead to problems, may lead to problems in some people but not in others, and may also lead to problems in some contexts but not in others.

3.5 Biological Facts and Social Values

If we cannot rely on scientific facts provided by statistics to decide on what is normal and what is abnormal, is concluding that health is an arbitrary norm that serves human preference the only way out? Sedgwick (1972) was probably the most notable proponent of this view. He states that sickness can only be established when we know about an alternative state of affairs that is more desirable:

> In the absence of this normative alternative, the presence of a particular bodily or subjective state will not in itself lead to an attribution of illness. Thus, where an entire community is by Western standards 'ill', because it has been infected for generations by parasites that diminish energy, illness will not be recognized in any individual except by outsiders. The Rockefeller Sanitary Commission on Hook-worm found in 1911 that this disease was regarded as part of normal health in some areas of North Africa (pp. 213–214).

Wakefield (1992, 2006) agrees with Sedgwick that whether a condition is undesirable is based on social values. It does not mean, however, that disorders can be reduced to values, because there are disvalued conditions that we would not think of as a disorder (e.g., poverty or being rejected sexually). Note that he assumes the reader agrees with him on what is to be considered a disorder and what not, prior to

[8] Canguilhem realised that to establish that something is wrong in physiology is not easy. He presents numerous examples of which urinary discharge is a particularly compelling one (1966/1989: 166–167). Chinese people between the ages of 18 and 25 release 0.5 cm^3 urine per minute (ranging from 0.2 and 0.7), whereas Europeans release twice that amount with oscillations between 0.8 and 1.5. These ranges do not even overlap. Canguilhem quotes Bernard (1865) "…a physiologist who took urine from the urinal at the train station through which people passed of all nations, and believed he could thus produce the analysis of average European urine" revealing that he was well aware of problems associated with averaging (1966/1989: 152).

his analysis of the concept. His lucid analysis of the concept of disorder leads him to propose the following definition:

> A condition is a disorder if and only if (a) the condition causes some harm or deprivation of benefit to the person as judged by the standards of the person's culture (the value criterion), and (b) the condition results from the inability of some internal mechanism to perform its natural function, wherein a natural function is an effect that is part of the evolutionary explanation of the existence and structure of the mechanism (the explanatory criterion) (Wakefield 1992: 384).

The important as well as most difficult concept in his view is 'natural function'. His analysis follows that of artefacts: a pencil's function is writing, the heart's function is to pump the blood around in the body, the function of language is communication, and the function of fear is useful for avoiding danger. These examples reveal a mechanistic view on (mental) functioning and may be justified by our knowledge of how people manage to survive and procreate. Proper functioning enhances longevity and fertility, which serves as the evolutionary aspect. Thus, talking to oneself as a means to remember a phone number is not a disorder, while 'hearing voices' is taken as a sign of schizophrenia. Note that people with schizophrenia have a reduced chance of offspring (MacCabe et al. 2009).

Although Wakefield's analysis has important similarities with that of Canguilhem, there are fundamental differences between them. They both adhere to the value orientation of disease or disorder. Unlike Wakefield, however, Canguilhem considers the relationship between the person/organism and the environment at the heart of his analysis on pathology rather than as an internal mechanism of the organism that went awry. Canguilhem's analysis reveals, as will be demonstrated below, a complex adaptive systems perspective, whereas Wakefield is a proponent of the mechanistic, information processing approach. I will follow the thoughts of Canguilhem rather than Wakefield for this, for reasons I will not explore further (see Varga 2011, on the untenability of Wakefield's natural function explanation).[9]

3.6 Normative and Pathological

Apart from Spicker (1987), the first substantial treatment of Canguilhem's work in the English-speaking scientific community was a special issue of the journal of 'Economy and Society'.[10] Although Canguilhem's discourse pertains mainly to physical diseases, it is equally applicable to mental disorders as shown by Margree (2002) and Verhoeff (2009).[11]

---

[9] Although Wakefield mentions the work of Foucault (1965), there is no reference to Canguilhem's analysis of the pathological in any of his works.

[10] The special issue of 'Economy and Society' (Osborne and Rose 1998) was the result of a conference held in London in 1996.

[11] I hasten to state that the distinction between the physical and the mental or psychological is just made for convenience. It remains to be seen whether it will hold up. An ontological distinction is certainly doubtful (see e.g., Sedgwick 1972; Wakefield, and many others in the journal of *Philosophy, Psychiatry, & Psychology*).

To understand Canguilhem, it is necessary to explain his view on science. He makes a fundamental distinction between the science of the non-living world (physics, mathematics, chemistry) and the life sciences (i.e., biology, the mother of the study of living organisms). Non-living sciences are value-free. It does not make sense to ask whether it is good or bad for a number to be a prime or whether redox reactions are healthy. Thirteen is a prime number because of the definition of primes and $4Fe + 3O_2 \rightarrow 2Fe_2O_3$ is a chemical reaction that describes the rusting of iron; both are simple value-free facts of reality. This is not the case with respect to biology.[12] In the living world, life may go wrong (Canguilhem 1966). Thus, organisms may be in situations that are good or promoting life and circumstances that are bad or detrimental to their survival.[13] The meaning or value for the organism in a given situation, that is, the organism-environment relationship, is a normative activity.

Healthy organisms are able to follow existing norms to maintain life given the existing circumstances. Healthy norms are dynamic (i.e., they change over time), because they display sufficient degrees of freedom to adapt to the environment and impose their normativity on the environment in return. Not being able to tolerate change is a sign of pathology, according to Canguilhem. Catching the flu and recovering from it is a sign of health. "Haemophilia is more of an anomaly than a disease" (Canguilhem 1966/1989: 140). Because people with haemophilia have to interact with the environment, this entails a risk of injuries, and thus the anomaly may turn into pathology.[14] With respect to psychiatric 'diseases', we have to ask ourselves whether a child who is restless and fidgety (conclusion based on being unable to sit still in her seat, roaming around the room, tapping and touching objects continuously) does in fact have ill health or to put it in psychological terms, suffers from Attention Deficit Hyperactivity Disorder (ADHD).

Recall that calling something unhealthy or a pathology implies that the organism is unable to adapt to change in order to maintain life. In my understanding, Canguilhem would never view ADHD-behaviour as pathological, because being restless and fidgety does not mean that one is unable to maintain life. ADHD-behaviour is another way of being, it is another order, but it is nothing more than that. Baron-Cohen (2008) has a similar view on autism and Asperger syndrome,

[12] Canguilhem (1966/1989: 136) nicely illustrates this as follows: "… the history of anomalies and teratology are a necessary chapter in the biological sciences, expressing the originality of these sciences – for there is no special science of chemical and physical anomalies…"

[13] Without getting into detail, I would like to raise a caveat about the fundamental difference between living and non-living things. The second law of thermodynamics states that the entropy of any isolated system not in thermal equilibrium almost always increases. In laymen's terms, physical systems have a natural tendency to become increasingly disordered. For example, gravel, the result of the erosion of rock, will not spontaneously return to rock again. For this to happen, energy has to be added to the system. Thus, rock and for that matter *all matter* is 'going wrong', because in the end all matter ceases to exist (cf., die). Note also Canguilhem's statement (1994: 117–118), "…one can still say that living things are systems whose improbable organization slows a universal process of evolution toward thermal equilibrium – that is, toward a more probable state, death."

[14] People with haemophilia have problems with the clotting of their blood, characterised by spontaneous bleeds or longer bleeding after an injury.

which he considers a *condition* rather than pathology (see also Hens in this volume on the issue of neurodiversity). People who are diagnosed with autism have a strong systemizing mind rather than the more prevalent empathizing mind (see also Grandin 2012).[15] A person who finds an environment that suits and promotes her behaviour is perfectly healthy. Because current practice in medical and social circles demands from people to conform to the statistical normal rather than to the biological normative, many children and adults are diagnosed with pathologies for conditions that Canguilhem would not consider to be a pathology at all (see, for a similar discussion, Giordano in this volume on healthy and pathological sexes).

Healthy organisms can also create new norms that enhance survival. A painter who loses both arms and learns to eat and draw with her toes has constituted a new norm for herself as an organism, one that is beneficial to her life. To still consider her to be in ill health is at odds with the analysis provided by Canguilhem. Witty Ticcy Ray in Sack's (1985) famous book 'The man who mistook his wife for a hat', is another example of someone who turns his condition, Tourette's syndrome, into the best of both worlds. During the week he takes Haldol to conform to normal, daily life, whereas in the weekends he is off the drug and uses his frivolous, frenetic, creative self to fully enjoy his musical gift as a jazz drummer. This way, Witty Ticcy Ray created alternative niches to lead a 'healthy' life.

Finally, healthy organisms can also change and impose new norms. Improving hygiene advances longevity. Enhancing longevity is a value judgment, because average life span is not a biologically normal given, but the socially normative. Another example is the change in smoking behaviour. Until the sixties, smoking was fully acceptable (at least for men). After science revealed the relationship between lung cancer and smoking, a steady decrease of smoking and its acceptance became apparent. Classical music used to be played in a tempo half of what it is now. At the end of the nineteenth century, new norms were successfully imposed by those with fast, technical brilliance (Wehmeyer in Honoré 2004). A current educational example is provided by the Finnish authority. The country's board of education states that typing is going to be far more useful than cursive writing, and it will therefore remove cursive writing from primary education's curriculum (Kerin 2015). Cursive writing used to be an important skill, hence a value, for professional development, but it is expected that being able to type fluently will be more important in the future. Again, good typing skills are not the biologically normal, but the socially normative.

Being able to create and impose new norms is also a sign of health.

> A living being is normal in any given environment insofar as it is the morphological and functional solution found by life as a response to the demands of the environment. Even if it is relatively rare, this living being is normal in terms of every other form from which it diverges, because in terms of those other forms it is normative... (Canguilhem 1966/1989: 144).

[15] Temple Grandin (2011) states that if autism had been eliminated, mankind would still be standing around in a cave, chatting and socializing and not getting anything done.

An anomaly (e.g., a mutation) can never in itself be pathological or normal; it expresses other potential norms for life. To establish normality in a laboratory cannot be a solution either, because it denies the fact that this is one of the many possible environments, one that is different from a natural situation (Bosman et al. 2013). Thus, nothing in itself can be pathological, because what is adaptive in one circumstance may prove to be maladaptive in another (see the current discussion on ADHD viewed from an evolutionary perspective, Matejcek 2003). In the next paragraph I will elucidate why Canguilhem's analysis of health and pathology is a clear example of a complex adaptive systems perspective.

3.7 Canguilhem and Complex Adaptive Systems Theory

A central and controversial concept in the work of Canguilhem is *vitalism*. This concept has been taken to mean that life is ontologically different from the non-living (see Greco 2005; Lecourt 1998; Verhoeff 2009 for a discussion). With Gutting (1989), I believe that this interpretation of Canguilhem's concept of vitalism is incorrect. Gutting's analysis paraphrasing Canguilhem reveals this quite clearly:

> In the sense that vitalism tried to reject the application of physics and chemistry to organisms, it was an obstacle to scientific progress. But, in another sense, vitalism was – and remains – a salutary reminder that, even if physico-chemical laws are fully applicable to organisms, vital phenomena still have distinctive features that exclude any facile reduction of them to inanimate systems (Gutting 1989: 41).

After all, "The laws of physics and chemistry do not vary according to health or disease" (Canguilhem 1966/1989: 220). Rather than accusing Canguilhem of holding an unscientific viewpoint, it seems more appropriate to argue that Canguilhem was way ahead of his time. In many parts of his work, we find clear manifestations of a *complex adaptive systems* thinker avant la lettre.

Complex adaptive systems (a concept first coined in 1984 by scientists from the Santa Fe Institute; Waldrop 1993) are systems such as organisms, immune systems, brains, insect colonies, and stock markets. These systems are able to learn and can adapt to changes in the (internal as well as external) environment. The properties or the behaviour of complex adaptive systems are emergent and cannot be deduced from the components that constitute the system. In his work published in 1943 Canguilhem acknowledges the characteristics of adaptation and emergence, when he talks about diabetes:

> …it is not a kidney disease because of glycosuria, nor a pancreatic disease because of hypoinsulinemia, nor a disease of the pituitary; it is a disease of an organism all of whose functions have changed […]. It seems very artificial to break up disease into symptoms or to consider its complications in the abstract. What is a symptom without context or background? What is a complication separated from what it complicates? When an isolated symptom or functional mechanism is termed pathological, one forgets that what makes them so is their inner relation in their indivisible totality of individual behavior (Canguilhem 1966/1989: 88).

Canguilhem was a great admirer of the psychiatrist and neurologist Kurt Goldstein (1878–1965) who stated that "If the organism is a whole and each section of it functions normally within that whole, then in the analytic experiment, which isolates the sections as it studies them, the properties and functions of any part must be modified by their isolation from the whole of the organism. Thus they cannot reveal the functions of these parts in normal life" (Goldstein 1940/1951: 10). This quote shows Goldstein's holistic approach to biology, which Canguilhem accepted as an important aspect of his view on pathology, organisms, and life in general.

A final example of Canguilhem's approach that is reminiscent of complex adaptive systems is the following: "An organism's behavior can be in continuity with previous behaviors and still be another behavior" (1966/1989: 87). Although there may be a continuous increase in quantity, this does not imply qualitative identity (see also Margree 2002). An example is water that is heated up. The kinetic energy that is transferred from the fire to the water increases continuously and does not show any visible change in the behaviour of the water between 65 and 66 until it changes from 99 to 100 degrees Celsius, despite the fact that the temperature in both cases only increased by 1 degree. The continuous change of the control parameter 'kinetic energy' causes a change in the behaviour variable after it crosses a threshold value. Water atoms in a liquid constitute an order different from the order in a gas. (Physical) Disease may occur when a pathogen has reached a particular, individual threshold. The order of the sick organism is qualitatively different from the order it had when it was healthy. Illness is therefore *order* and not disorder. Being ill is being different from being healthy. Different orders have different norms and that is why these conditions are value-laden.

3.8 Pathos, Values and Responsibilities

Being ill or having a disorder implies *pathos* according to Canguilhem, it is "... the direct and concrete feeling of suffering and impotence, the feeling of life gone wrong" (Canguilhem 1966/1989: 137; italics in original). In a quote of Leriche he emphasizes the *suffering* aspect: "disease is what irritates men in the normal course of their lives and work, and above all, what makes them suffer" (Canguilhem 1966/1989: 91). Hence, pathos or suffering is an important, albeit not sufficient, aspect of pathology.[16] Canguilhem agrees with Goldstein that a pathological norm is above all an individual norm. This implies that it is mostly the individual who decides whether she is ill or has regained health pertaining to somatic and psychological anomalies alike.

Although the pathological norm is an individual one, it does *not* mean that pathology is localised *in* the individual (or *in* the environment for that matter). Pathology, as explained above, is the diminished capacity of the organism in that

[16] Suffering is not a sufficient condition, because a healthy person may suffer during an illness. Being ill is not necessarily a pathology; recovering from an illness is a sign of health.

particular environment to conform to existing norms (i.e., to make life profitable for survival or pleasurable). In other words, pathology emerges when organism A in environment B is unable to conform to norm C. Because pathology is an emergent property of the relationship between the organism and its environment, a pathology or disorder cannot be assigned to a characteristic of the organism *or* the environment. An emergent property[17] means that the property cannot be reduced to the assumed components that gave rise to it. With respect to pathology and disorder this can only lead to one conclusion, namely, that it is incorrect to state that person A has disorder X.

If the above makes sense then it will not come as a surprise that Canguilhem does *not* consider medicine to be a science. Physics is a science, but medicine, like psychiatry, or clinical (child) psychology (also known as orthopedagogy in the Netherlands) is not, because the client's[18] experience, that is, suffering, cannot be reduced to scientific objectivity. In Canguilhem's words: "... the life of the living being, were it that of an amoeba, recognizes the categories of health and disease only on the level of experience, which is primarily a test in the affective sense of the word, and not on the level of science. Science explains experience but it does not for that annul it" (Canguilhem 1994: 356). He views medicine,[19] and for that matter all disciplines concerned with restoring 'health', as "...a technique or art at the crossroads of several sciences..." (Canguilhem 1966/1989: 34). Clinical practice will never be a science, despite the fact that its methods are effective as a result of science, because science cannot dictate norms to life (Trnka 2003). To be more concrete, the goal in medicine is to provide a therapy or cure to diminish suffering or restore health. Despite the 'scientific fact[20]' that a daily dose of morphine alleviates many pains rather effectively, doctors will not use this type of medicine unless there is no other option, because of its highly addictive property. In clinical practice, a doctor needs to weigh the pros (pain reduction) and cons (drug dependency) when prescribing a medicine. Because pros and cons represent different norms, science can never provide the answer.

Assessing suffering and providing treatment are the prime goals of the clinical disciplines. Current practice of the assessment of an individual's problem requires diagnostic tools. As I showed earlier, most tests in the medical discipline, such as thermometers and sphygmomanometers, are valid tools, whereas psychological tests are not. Note that despite the use of valid instruments, diagnoses in medicine are nevertheless value-laden. It is we who decide that a certain blood pressure needs

[17] O'Connor and Wong (2012) use a generic definition: "Emergent entities (properties or substances) 'arise' out of more fundamental entities and yet are 'novel' or 'irreducible' with respect to them."

[18] In all of psychology (and also sometimes in psychiatry) the term client is used rather than patient, contrary to the usual choice of terms in medicine. It shows, and that is another interesting aspect, that somehow being *psychological ill* does not mean that one is a patient as in being *physically ill*.

[19] Medicine means 'the healing art' (www.etymonline.com).

[20] A scientific fact indicates here that it has not only been proven that opium has this capacity (something we knew all along), scientists also claim to have found the mechanism of how the substance works in the body.

to be treated, because its level is considered unacceptable. An unacceptable level is a value statement, irrespective of whatever (good) reason. Apart from the fact that psychological tests are not valid, because they lack all of the requirements put forward by Borsboom et al. (2004), further analyses revealed that all psychological tests have normativity built-in.

I have also explained that taking a measurement establishes a relationship between what measures and what is measured. In terms of psychology, the test, the diagnostic, and the 'client' constitute an irreducible whole. Any evaluation (i.e., diagnosis) is always the result of the relationship between client and clinician (potentially with measurement tools). Assessment is an intervention and this intervention changes the relationship of the client with the environment (which includes the diagnostic). A clinician cannot withdraw from this relationship to assess the client's behaviour/suffering. This is not just practically impossible (someone has to interact to establish something), it is also practically unavoidable. The relationship between organism and environment provides the stage for diagnosis and treatment and this relationship is inherently normative.

If a disorder, preferably *different* order[21] (for reasons explained above), is an emergent phenomenon, it calls for a description in terms of the relationship between the individual and her circumstances. This means that those who are dealing with people in need should appraise the relationship rather than its components, the organism or the environment. Hence, simply checking characteristics listed in manuals like the DSM does not lead to a 'valid' evaluation of the situation, because of the emphasis on the individual rather than on the relationship. Canguilhem's analysis of pathology or disorders provides people who look after children, such as parents and other caretakers and those who are engaged in the wellbeing of individuals in need, with a huge responsibility. This responsibility requires a thorough reflection of the situation of people who suffer.

What does my analysis entail for the current situation in many countries in which there is no help without an official DSM-diagnosis? People who suffer, but whose suffering cannot be reduced to one of the many official labels, will not be eligible for financial compensation even when they feel that treatment is required. This policy is to a large extent the responsibility of the scientific community who made policy makers believe that we can truly distinguish between the normal and the not-normal. One does not need to dig deep to see that the DSM is not based on scientific facts, but on politics. Interest groups whose suffering was not officially recognized have sometimes lobbied successfully for recognition (e.g., post-traumatic stress syndrome by Vietnam veterans; DSM-III), whereas other interest groups were successful in removing a disorder from the DSM (e.g., homosexuality was fully discarded from the DSM-IV). A quote from Lucy Johnstone's book (2000, see Rowe 2010) endorses my analysis, "To admit the central role of value judgments and

[21] In a similar vein is the suggestion of Bettinger (2015) who uses the term 'signal' behaviour rather than 'problem' behaviour. The former is not just less evaluative, it also refers to communication that underlies all (human) behaviour, which in turn emphasizes the relationship between client and caretaker.

cultural norms [in the creation of the DSM] is to give the whole game away. The DSM has to be seen as reliable and valid, or the whole enterprise of medial psychiatry collapses."

Because the DSM cannot provide a valid diagnosis, its use only provides a pseudo-ontological status for which the scientific community has to take full responsibility. Parents, caretakers, and clinicians need to recognize that suffering is at the heart of the problem, not being different. The fact that a child or adult has the DSM-diagnosis 'Autism Spectrum Disorder', which is viewed as a lifelong affliction, provides the individual support and financial compensation. But, someone who is diagnosed with autism may or may not suffer, just like someone who contracted poliomyelitis may suffer or not.

A person in need is a person in need in a specific time and place, with its unique history in relationships with significant others and the clinician (i.e., circumstances). And, because circumstances as well as the organism are continuously changing (life is not static, otherwise it would not be life), the needs of care-dependent people require continuous attention, that is, monitoring over time. But above all, they require a parent, caretaker, and clinician to always reflect on the norms that come with our perceptions of (diagnoses), and actions on (interventions), care-dependent people. Scientists and politicians may need to reconsider the established values of health and disease.

References

Achenbach, T.M. 1991. *Manual for the child behavior checklist/4-18 and 1991 profile*. Burlington: University of Vermont, Department of Psychiatry.

American Psychiatric Association. 2013. *Diagnostic and statistical manual of mental disorders*, 5th ed. Washington, DC: Author.

Baron-Cohen, S. 2008. *Autism and Asperger syndrome. The facts*. Oxford: Oxford University Press.

Bettinger, G. 2015. *Door stil te staan kom je verder [You travel further by standing still]*. Amsterdam: SWP.

Borsboom, D., G.J. Mellenbergh, and J. van Heerden. 2004. The concept of validity. *Psychological Review* 111: 1061–1071.

Bosman, A.M.T., R.F.A. Cox, F. Hasselman, and M.L. Wijnants. 2013. From the role of context to the measurement problem: The Dutch connection pays tribute to Guy Van Orden. *Ecological Psychology* 25: 240–247.

British Psychological Society. 2011. *Response to the American Psychiatric Association DSM-5 development*. Retrieved January 6, 2015, from apps.bps.org.uk/_publicationfiles/consultation-responses/DSM-5%202011%20-%20BPS%20response.pdf.

Canguilhem, G. 1966. *The normal and the pathological*. New York: Zone Books.

Canguilhem, G. 1994. *A vital rationalist. Selected writings from Georges Canguilhem*, ed. François Delaporte. New York: Zone Books.

Decker, H.S. 2010. *A moment of crisis in the history of American psychiatry*. historypsychiatry.wordpress.com/2010/04/27/a-moment-of-crisis-in-the-history-of-american-psychiatry.

Foucault, M. 1965. *Madness and civilization: A history of insanity in the age of reason*. Trans. R. Howard. New York: Pantheon. (Original work published in 1964).

Frances, A.J. 2012. *DSM 5 Is guide not bible — Ignore its ten worst changes*. Retrieved January 6, 2015, from www.psychologytoday.com/blog/dsm5-in-distress/201212/dsm-5-is-guide-not-bible-ignore-its-ten-worst-changes.

Frances, A.J., and T. Widiger. 2012. Psychiatric diagnosis: Lessons from the DSM-IV past and cautions for the DSM-5 future. *Annual Review of Clinical Psychology* 8: 109–130.

Goldstein, K. 1940. *Human nature in the light of psychopathology*. Cambridge, MA: Harvard University Press.

Grandin, T. 2011. *The way I see it: A personal look at autism and Asperger's*. Revised and expanded 2nd edition. Arlington: Future Horizons.

Grandin, T. 2012. *Different…not less*. Arlington: Future Horizons.

Greco, M. 2005. On the vitality of vitalism. *Theory, Culture & Society* 22(1): 15–27.

Greenfield, L. 2013a. *The real trouble with DSM-5. Are its problems hereditary?* Retrieved January 8, 2015, from www.psychologytoday.com/blog/the-modern-mind/201305/the-real-trouble-dsm-5.

Greenfield, L. 2013b. *Mind, modernity, madness: The impact of culture on human experience*. Cambridge, MA: Harvard University Press.

Grohol, J.M. 2013. *Did the NIMH withdraw support for the DSM-5? No*. Retrieved January 6, 2015, from psychcentral.com/blog/archives/2013/05/07/did-the-nimh-withdraw-support-for-the-dsm-5-no/.

Gutting, G. 1989. *Michel Foucault's archaeology of scientific reason*. Cambridge: Cambridge University Press.

Hacking, I. 2013. Lost in the forest. *London Review of Books* 35(15): 7–8.

Harper, D. n.d.. *Online etymology dictionary*. Retrieved on May 24, 2014, http://www.etymonline.com.

Honoré, C. 2004. *In praise of slowness: How a worldwide movement in challenging the cult of speed*. New York: HarperCollins Publishers.

Johnstone, L. 2000. *Users and abusers of psychiatry: A critical look at psychiatric practice*. London: Routledge.

Kaufman, A.S. 2009. *IQ testing 101*. New York: Springer.

Kerin, L. 2015. *Finland scraps cursive writing lessons, sparking discussion over future of handwriting in classrooms*. Retrieved April 19, 2015, from www.abc.net.au/news/2015-02-03/finland-scraps-cursive-writing-lessons-in-schools/6066826.

Lecourt, D. 1998. Georges Canguilhem on the question of the individual. *Economy and Society* 27: 217–224.

MacCabe, J.H., I. Koupil, and D.A. Leon. 2009. Lifetime reproductive output over two generations in patients with psychosis and their unaffected siblings: The Uppsala 1915–1929 birth cohort multigenerational study. *Psychological Medicine* 39: 1667–1676.

Margree, V. 2002. Normal and abnormal: Georges Canguilhem and the question of mental pathology. *Philosophy, Psychiatry & Psychology* 9: 299–312.

Matejcek, Z. 2003. Is ADHD adaptive or non-adaptive behavior? *Neuroendocrinology Letters*, 24(3/4):148–150. pii: NEL243403L01.

O'Connor, T., and H.Y. Wong. 2012. Emergent properties. *The Stanford encyclopedia of philosophy*. plato.stanford.edu/archives/spr2012/entries/properties-emergent.

Osborne, T., and N. Rose. 1998. Introduction. *Economy and Society* 27: 151–153.

Rowe, D. 2010. *Mental health diagnoses mask the real problems*. Retrieved April 26, 2015, from www.theguardian.com/commentisfree/2010/jul/29/mental-health-diagnostic-manual.

Sacks, O. 1985. *The man who mistook his wife for a hat*. London: Picador.

Sedgwick, P. 1972. Mental illness is illness. *Salmagundi* 20: 196–224.

Seeley, K. 2004. Gifted and talented students at risk. *Focus on Exceptional Children* 37(4): 1–8.

SPD Foundation. 2012. *Final decision for DSM-V*. http://spdfoundation.net/dsmv1.html.

Spicker, S.F. 1987. An introduction to the medical epistemology of Georges Canguilhem: Moving beyond Michel Foucault. *The Journal of Medicine and Philosophy* 12: 397–411.

Trnka, P. 2003. Subjectivity and values in medicine: The case of Canguilhem. *Journal of Medicine and Philosophy* 28: 427–446.
van Os, J. 2014. *De DSM-5 voorbij! Persoonlijke diagnostiek in een nieuwe GGZ* [Past the DSM-5. Personalised diagnostics in a new Health Care System]. Leusden: Diagnosis Uitgevers.
Varga, S. 2011. Defining mental disorder. Exploring the 'natural function' approach. *Philosophy, Ethics, and Humanities in Medicine* 6:1.
Verhoeff, B. 2009. *Georges Canguilhem and the concept of autism. Pathology at the border of the vital and the social.* Master thesis, University of Amsterdam. http://dare.uva.nl/cgi/arno/show.cgi?fid=164313.
Wakefield, J.C. 1992. The concept of mental disorder: On the boundary between biological facts and social values. *American Psychologist* 47: 373–388.
Wakefield, J.C. 2006. What makes a mental disorder mental? *Philosophy, Psychiatry, & Psychology* 13: 123–131.
Waldrop, M.M. 1993. *Complexity: The emerging science at the edge of order and chaos*. New York: Simon and Schuster.

Chapter 4
Advances in Genetics and Neuroscience: A Challenge for Personalizing Child and Youth Health Care?

Elena V. Syurina and Frans J.M. Feron

Abstract Child and Youth Health Care (CYHC) is the branch of primary care that aims at an early identification of paediatric health problems with a special focus on developmental conditions. Recent advances in genetics and neuroscience reveal that developmental conditions are in most cases multifactorial: they do not have one single cause but rather develop due to longitudinal interplay between biological and environmental factors. In order to incorporate new scientific knowledge into the daily practice of CYHC important changes are necessary. The CYHC system should move away from a purely diagnostic approach and instead focus on the early identification of children at risk, the prevention of developmental problems at various stages and the provision of personalised interventions. To that end, it should follow the principles of personalised care, adapt its interventions to each individual child and work together with parents. This shift also calls for a more active involvement of parents within CYHC.

Keywords Primary Paediatric Health Care • Child and Youth Health Care • Genetics • Neuroscience • Personalised Health Care • Prevention • Mental Health

E.V. Syurina (✉)
Athena Institute for Research on Innovation and Communication in Health and Life Sciences, Faculty of Earth and Life Sciences, Vrije Universiteit, Amsterdam, The Netherlands

Department of Health, Ethics and Society, School for Public Health and Primary Care (CAPHRI), Maastricht University,
Maastricht, The Netherlands
e-mail: e.syurina@maastrichtuniversity.nl

F.J.M. Feron
Bioethics Institute Ghent, Department of Social Medicine, School for Public Health and Primary Care (CAPHRI), Maastricht University, Maastricht, The Netherlands
e-mail: f.feron@maastrichtuniversity.nl

K. Hens et al. (eds.), *Parental Responsibility in the Context of Neuroscience and Genetics*, International Library of Ethics, Law, and the New Medicine 69,
DOI 10.1007/978-3-319-42834-5_4

4.1 Introduction[1]

Children are a special population group in our society. This is not only because of biological reasons, such as their rapid growth and development, but also for social reasons. They have limited (yet evolving) capacities to make informed decisions about own interests and are in need of adult supervision (Smidt 2013). When children face challenges in their growth or development, it is of utmost importance that they receive proper support or treatment. Childhood problems that remain untreated not only cause immediate suffering, but can also have life-long negative consequences for the children themselves, their families and the wider society (Goodman et al. 2011). For this reason, some have argued that good preventive and pre-emptive health care provisioning is needed as it will result in long-lasting benefits both for individual children and for society in general (Kieling et al. 2011). In this chapter, we will discuss how recent findings from genetics and neuroscience could and should be integrated into established Child and Youth Health Care (CYHC) systems to optimise the care for young people. We will argue that this requires a paradigm shift. Instead of focusing on diagnosing established developmental problems, professionals should monitor children before the onset of clinical problems and provide effective preventive interventions during all phases of emerging disturbances.

In order to set the scene, we will first present a general overview of CYHC systems. Why did these systems appear? How did they develop and what is their current status? These systems can and do vary among different countries. As a practical example, we will present the Dutch system. Given that the main focus of CYHC is the early identification and diagnosis of childhood developmental problems, we will then provide a short insight into various bio-psychosocial determinants of development. What is development? How does it occur? What biomedical and environmental factors play crucial roles in this process? We will also provide a brief overview of recent advances in the fields of neuroscience and genetics on normal and abnormal brain development and will discuss how these could, and in our view also should, influence current CYHC practice.

One important aspect of current advances in neuroscience and genetics is a fundamental change in the perspective of how child development occurs. Prior to this change, the mainstream approach to child development was causal: it held that development takes place on the basis of single factors and problems occur in case of distinct adversities. Currently, a more integrative and multi-factorial model is being proposed (Snyderman and Langheier 2006). That is, in case a child does indeed experience developmental difficulties, this is no longer considered to be a state that is caused by some single factor. Instead, it is perceived as a situation that develops step by step and is due to a variety of biological as well as environmental influences. Moreover, initially occurring developmental difficulties can temporarily or even permanently go into remission (Greenberg et al. 2003). This new perspective on

[1]We would like to thank the editors of this volume for their assistance in improving this chapter, especially Dr. Dorothee Horstkötter.

child development and developmental disturbances calls for major adjustments of the CYHC system. It is not only that recent findings from genetics and neuroscience should be taken into account, but also that it is of paramount importance that the involvement of parents in CYHC is increased. The parents, however, cannot do it alone; the system has a responsibility to provide opportunities, information and support. Thus, in the last part of our chapter we will discuss how these aims can be achieved.

4.2 Historical Insight in Child and Youth Health Care

Even though child health has been a matter of interest for several centuries, the care systems that monitor it developed at different rates in various countries. The focus of these systems is not static, but constantly evolving to adapt to the changing societal situation and demands. In the eighteenth and nineteenth centuries in Europe childhood mortality and morbidity were much higher than they are today. Due to the lack of clean water supplies, poor waste disposal systems and overcrowding in cities and towns, infectious diseases such as tuberculosis, cholera and influenza were among the major causes of death among child and adult population (Fridlizius 1989). In the nineteenth century, whooping cough, diphtheria, poliomyelitis and smallpox were among the most prevalent childhood diseases (Knodel 2002). Through the development of personal and household hygiene rules, food protection and preservation, which were later followed by the introduction of fresh water supplies, national vaccination programs and other public health actions, there was a drastic decline in childhood mortality and morbidity (Cutler and Meara 2001; Cutler and Miller 2005). While some of these issues still play an important role in developing countries (and some European ones), the priorities of CYHC in most Western states have changed fundamentally. Today, the focus is mainly on developmental disturbances and disorders, psychosocial issues, child abuse and neglect, juvenile substance abuse as well as the care for chronically ill children and childhood obesity (Coker et al. 2013; Hagan et al. 2008).

In the Netherlands, the CYHC system is closely embedded in general health care. It is organised by means of a special law, the Public Health Act (Wet Publieke Gezondheid – WPG), adopted in 2008 (Dute 2008). CYHC deals with prevention and early diagnosis of diseases as well as identification of risks that may interfere with development and daily functioning of the child (Boomsma and Pijpers 2008). All children from birth until the age of 18 living in the Netherlands are entitled to get support from CYHC services, which is provided free of charge.

The system is currently divided into two branches: Mother and Child Clinics (MCC) support children aged 0–4 and Child and Youth Healthcare Departments of the Regional Public Health Services (RPHS) provide services to children aged 4–18. During regular consultations at the MCC the health, development and growth of very young children is monitored and parents can pose questions regarding the upbringing of their children. Moreover, parents are given advice on nutrition, care and prevention: e.g., what position is best for sleeping, how to minimise risks of common colds etc. At the age of four, the preventive care is taken over by the RPHS. The tasks of RPHS are similar to those of MCC. Children's and juveniles' growth is monitored (by the use of such measures as body mass and height), their social development is surveyed (via various questionnaires) and support and relevant interventions are offered if developmental problems are identified. Even though visits to RPHS care are less frequent than in MCCs, parents can make an appointment at any time when they are concerned about their child's development.

In the past years, the focus of CYHC has shifted from detection and treatment of somatic disorders [which today, at least in the Netherlands, is mainly the task of general practitioners] to the identification of children at risk of developmental problems such as behavioural or emotional difficulties, learning disabilities or other issues interfering with a child's daily functioning. To this end, CYHC is also increasingly investigating the social environment the child lives in. Correspondingly, whenever problems are identified, these are approached by addressing children's families and/or their school environment. This focus on psycho-social difficulties, as opposed to the more somatic complaints, also provides for a clear distinction between the responsibilities of general practitioner and CYHC physician and thereby helps avoiding the duplication of tasks. Therefore, CYHC physicians do not provide curative help. If the symptoms of the child cluster to DSM standards for mental health diagnosis (American Psychiatric Association 2013), the CYHC physician refers the child to a specialist.

4.3 Child Development as the Current Target Area of CYHC

Child development is generally described as the combination of biological, psychological and emotional changes that occur between birth and the end of adolescence, as the individual progresses from dependency to increasing autonomy (Kail 2001). Besides purely biological development, such as growth and maturation, research has always paid much attention to the psycho-social development of children. Several theories have been advanced to describe and explain this development: psychoanalytical, biological, cognitive etc. Piaget, for example, focused on intellectual or cognitive development and tried to formulate a theory on repeating and age-dependent developmental stages present in all children (Piaget and Cook 1952). In addition, Skinner developed what he called the theory of operant conditioning, according to which children's behaviour can be simplified to being a result of positive and negative reinforcements (McLeod 2007).

The aim of these and several other theories was to describe typical development of children in order to differentiate between normal, borderline and abnormal ranges. Today, a variety of screening tools are being used in CYHC to monitor child development. The majority of them share the same goal: to help physicians decide whether a specific child's development shows merely a slight variation of what is actually a normal pattern, or whether the child is exhibiting early signs of developmental difficulties. Some tools provide a longitudinal assessment of the developmental pathway of the child. The Peabody Developmental scales (Folio and Fewell 1974) and the Baley Scale of Infant Development (Bayley 1993), for example, assess various aspects of development, including motor, cognitive and language development. Others focus on assessing the child's present behaviour and indicate whether parents and physicians should take further actions at any specific moment. The Strengths and Difficulties Questionnaire (SDQ) (Goodman 2001) and the CBCL (Child Behaviour Checklist) (Achenbach and Edelbrock 1981) investigate, for example, whether a child currently has symptoms of emotional problems or observable signs of attention difficulties.

Psycho-social screening is considered a reliable method to identify and cluster existing difficulties. However, it focuses purely on outer symptoms and the visible behaviour of the child, and a child's development is much broader than that. It operates on several levels covering genetics, biology, psychology and societal issues and only some of the features relevant in these regards are directly observable. In order to be able to effectively use behavioural and symptom information, physicians must also be well informed about underlying biological processes and mechanisms and their causes, e.g. genetic or neurobiological causes that may lead to the development of a condition. Recent technological advances are allowing us to move from solely observing and analysing the behaviour of the child, towards broader investigation of the underlying brain processes.

4.4 Child Development: Psychological, Environmental, Neurobiological and Genetic Aspects

One of the most important parts of childhood development is the development of the brain. It creates the basis for a child's personality, behaviour and all learning processes. Brain formation is pre-programmed at conception and continues throughout pregnancy, infancy and childhood into adolescence and beyond. It is usually divided into two distinct phases: pre-natal and post-natal. Pre-natal brain development is complex, involving multiple processes, often occurring simultaneously, that can be influenced by the interaction between genetic and environmental factors. For example, maternal stress or substance abuse typically have negative effects on foetal brain development, and can lead to foetal alcohol spectrum disorders (Raznahan et al. 2012). Yet, the developing child's genetic susceptibility can either aggravate or mitigate these effects (Mason and Zhou 2015). In general, by the end of the pregnancy, all major brain regions are formed and important bodily functions (such as moving and breathing) are already regulated by the brain.

One of the underlying processes of after-birth development in childhood is learning. Very young children learn to distinguish sounds (which is also a basis for language formation), recognise faces and emotions and many other features (Bee and Boyd 2000). Learning leads to the formation of a large amount of synapses:

connections between individual neurons. This process is called synaptogenesis. It is known that synaptogenesis continues throughout the life of every healthy individual, but an explosion of the synapse formation can be observed during critical stages of development. For instance, the development of emotional control skills peaks around the age of one and starts slowly declining after two, while development of understanding of the meaning of symbols peaks slightly later, around the age of two (Illingworth 2013; Birch et al. 2013). Together with synapse formation, the opposite process of elimination of the synapses occurs. This process is called synaptic pruning and entails the deletion of unnecessary or less effective connections while strengthening the most useful and important ones. Synaptogenesis and synaptic pruning are the results of genetically predetermined processes, but are also highly influenced by the environment in which the child lives. For instance, the negative long-term health consequences of child abuse have been known for some time. People abused as children are at higher risk of developing mental disorders, using drugs, attempting suicide, acquiring sexually transmitted infections, or engaging in risky sexual behaviour (Norman et al. 2012). More recently, however, biological explanations for this correlation have been discovered: in situations of extreme neglect, where children do not get enough brain stimulation, a dramatic decline in synaptogenesis and synaptic pruning has been observed (Twardosz and Lutzker 2010). Moreover, the effects of neglect are not limited to synaptic processes: neglect also damages the integrity of white matter, a process that in turn leads to much less effective cognitive functioning (Hanson et al. 2013).

Unfortunately, effective methods of monitoring the specific biological processes of brain development are not yet available in everyday clinical practice, thus the most professionals can do is to review the environmental factors that a child is exposed to (Ben-Arieh et al. 2013). However, when doing so one should be aware that the influence of these effects is not linear: at some intervals during childhood, it can have more detrimental or supporting effects than at others (Knudsen 2004). Often these periods are linked to major changes in the child's life: development from infant to toddler, from toddler to pre-school child and further (Andersen 2003). Furthermore, critical periods for various functions do not occur at the same time: peer social skills development is boosted around the third year (Shonkoff and Meisels 2000), while adolescence is a critical period for the development of impulse and emotion control (Crews et al. 2007; Steinberg 2007). Environment is not the only factor that influences the child's brain development: throughout life, but especially in the early stages, much depends on genes and their interaction with the environment. Some 50–60 % of all known genes contribute to brain development in some way, and genetic mechanisms have been shown to be extraordinarily important for these processes (Davidson 2012), affecting, for example, white matter integrity (Chiang et al. 2011). Another example is the mutation in the aristaless-related homeobox gene that was strongly linked to the condition called Agenesis of the corpus callosum (white brain matter connecting the two brain hemispheres, which in these cases is partly or completely absent) (Paul et al. 2007).

However, the more we learn about genetics, the stronger the evidence that the cases of single gene mutations affecting brain development are quite rare. Most of the processes leading to development of various developmental conditions are of an epigenetic nature. Epigenetics aims to explain and map alternations in the gene functions and cell potential that cannot be explained by changes in DNA. Attention has also been paid to the mutual influence of environment and genes. For example, the genetic component in Attention Deficit Hyperactivity Disorder (ADHD) seems to be quite strong (Martel et al. 2011), but no single gene has been detected that would cause the condition by itself (Archer et al. 2011). Recent findings suggest that epigenetic factors also play a role (van Mil et al. 2014). Nonetheless, simplistic concepts of using genome information to predict a person's health status development have proved to be unrealistic (Sawicki et al. 1993). Even a full DNA analysis cannot predict which behavioural disturbances a child will develop or at which age. Even though there is increasing evidence that most common conditions have a genetic component, environmental effects on the genome throughout life can influence whether conditions manifest or remain dormant (Allis et al. 2007).

Hence, it makes no sense to consider genetics or environmental causes of common childhood conditions as separate factors: instead, we need to discuss the interplay of the two. This interaction is even more complex than was originally assumed: it was recently discovered that epigenetic mechanisms can also cause transgenerational changes in the genome of the individual that will be passed to her offspring (Youngson and Whitelaw 2008). For instance, extreme early life stress can cause persistent changes in the methylation of *BDNF* DNA causing altered *BDNF* gene expression in the adult prefrontal cortex, not only in mothers experiencing early life stress, but also in their children (Roth et al. 2009). One of the potential effects of these changes is the effect on behaviour in stressful situations.

However, despite the fact that knowledge of genetics, epigenetics and neuroscience is increasing in hospital-based care, its use in the CYHC remains very limited. This is unfortunate as this body of knowledge can contribute to the development of the new efficient and effective interventions. Below we will suggest ways in which this can be changed.

4.5 "Growing into Deficit": A New Paradigm for CYHC

As discussed in the previous paragraph, recent findings have brought about a paradigm shift in our understanding of the development of psycho-social and mental health related problems. Rather than being caused by a single factor (be it either genetic or environmental), developmental problems instead emerge as a consequence of the complex interaction of various genetic, neurobiological, psychological and environmental factors. These factors mutually influence each other and can accumulate with time. This means that children who show developmental conditions do not do so suddenly. While at first being hardly present or recognizable, over time the influence of the various factors builds up (causing an increasing number and severity of the symptoms) until a certain behavioural threshold is reached and a DSM diagnosis applies. In some cases, due to, for example, positive environmental

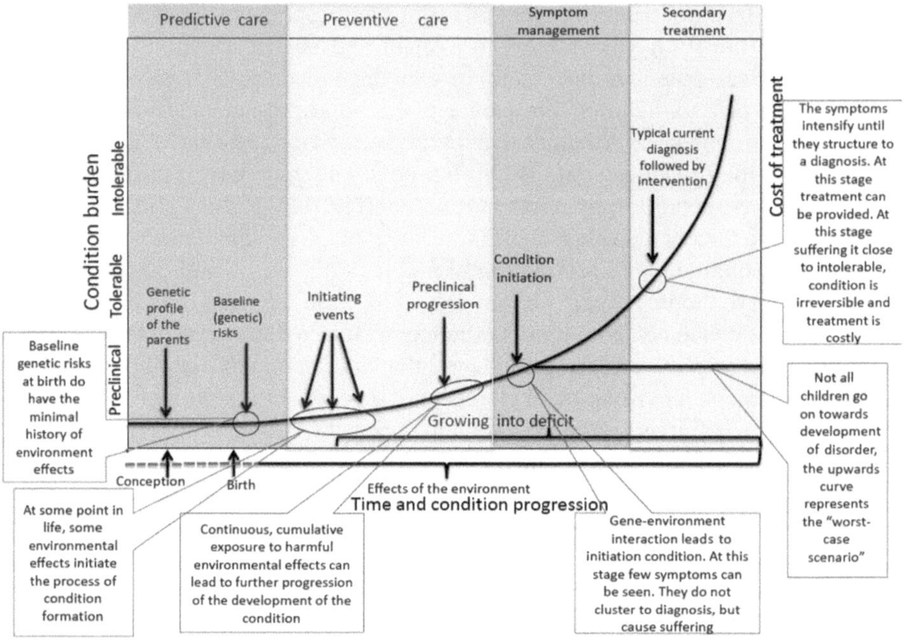

Fig. 4.1 Adaptation of Snyderman & Langheier's curve representing 'growing into deficit' and developing common complex conditions

influences, the threshold is never reached, and a child does not develop a full-fledged mental disorder. Developmental problems hence are to be characterized as both gradual and additive. Elsewhere (Syurina et al. 2013), we established the 'growing into deficit' developmental model which we adapted from Snyderman and Langheier (Snyderman and Langheier 2006) and are here refining further. This model represents child development as a curve (Fig. 4.1). The curve presented here associates children's lifetime and condition progression with the increasing burden the condition imposes on the child.

In the 'growing into deficit' developmental model, a curve represents the course of development of the majority of complex conditions with special attention to developmental abnormalities such as autism spectrum disorders (ASD), ADHD, learning disabilities etc. It does not apply to monogenetic conditions, which are exclusively the result of single gene mutations (such as cystic fibrosis or neurofibromatosis) nor to environmentally induced diseases, which are caused by pathogens outside the body, such as viral or bacterial infections. Every child is born with a certain genetic profile. This profile renders children more or less susceptible to the development of certain problems, meaning that some (genetically susceptible) children are particularly at risk in case they experience adverse environmental features. Others that possess a more 'robust' genetic profile might develop in more favourable ways despite similar environmental adversities. Moreover, as discussed before and as described by Hens in this volume, recent findings in epigenetics indicate that the expression of genes can also be influenced by the environment a child grows up and lives in.

So according to our model, a child's genetic profile forms the basis for the future. At some point in life, negative life events or adverse environmental exposures can serve as initiating events for a developmental problem. At an early stage, however, the progression of a condition may still be temporarily or permanently stopped. The curve can take any form, from remaining flat, which means that no significant burden of diseases develops, to an exponential growth, which means that the burden rises continuously and increasingly fast. That is, if negative effects continue to accumulate, what had been preclinical in first instance might develop further and a clinical stage might be reached. The symptoms get so burdensome that it becomes very hard for children to cope. This in turn has direct negative effects on the quality of life of a child's whole family (it can cause tension and arguments, prevent the child from participating in school activities etc). However, the observable symptoms and difficulties do not - yet - cluster to a definitive DSM diagnosis. In our model, this is the phase between the initiation of any condition and the stage of official diagnosis that is called 'symptom management'. Given that every child who develops a diagnosable mental health problem first goes through this preclinical phase, we consider it plausible that children can already profit at this stage from preventive and therapeutic interventions, ideally resulting in a situation where fewer children develop diagnosable mental health disorders. For instance, a recent study showed the positive effects of early-stage interventions targeting infants at risk of autism (Green et al. 2015). We would like to emphasise that it is of crucial importance that the CYHC and other paediatric mental healthcare systems acknowledge this situation and offer care and support to children before their problems and symptoms cluster into a recognized pattern of an official diagnosis. This might be an important step to control and minimize the burden of the symptoms and accompanying suffering. Unlike the system that is currently in place, our model proposes that one should focus on the burden that a certain cluster of symptoms imposes on children and families, independent of whether this cluster amounts to a full-fledged diagnosis. This will allow for a much more flexible and personalized approach.

4.6 Towards a Sustainable CYHC

The CYHC system should be able to screen and monitor children's health, growth and development, to make longitudinal assessments and to use these data to improve children's health outcomes and to prevent developmental problems. Recurrent assessments and screenings can also be a good basis to form good, reliable and trusting relationships between children, their parents and CYHC health professionals. Currently, however, the system makes sub-optimal use of its potentials. It is rigidly focused on official diagnoses, frequently does not take recent scientific findings into account, and fails to build up relationships with its main clientele: parents and their children. In the following, we will present current weaknesses of the system in greater detail and make concrete suggestions on how these weaknesses can be strengthened and how the system can be rebuilt in a more integrative and sustainable manner.

4.6.1 Excessive Focus on Diagnosis

The current system's focus on official diagnoses is problematic in several ways. First, it ignores children whose symptoms are troublesome already but whose issues do not – yet – cluster into a diagnosis. Second, given that most mental health diagnosis face serious stigmatization, people often want to avoid the system, are reluctant in asking for help and muddle through until the burden becomes unbearable.

The prevalence of children with significant impairments due to developmental problems, but without a definitive diagnosis, is estimated to be twice that of children who meet the internationally recognized DSM criteria for mental health disorders (Committee on Psychosocial Aspects of Child Family Health Task Force on Mental Health 2009). Because of their impairments, these children cannot live up to their full potential. Nonetheless, intervention options for them are limited and difficult to obtain. However, by the time diagnosis becomes possible, much suffering has already occurred and preventive interventions are no longer applicable. Instead, specialist-based curative treatment and care must be offered. Such treatments have a series of drawbacks compared to preventive interventions at the earlier stages. They are usually longitudinal and may take years to yield the desired effect. They require much effort from the child and her family and are often difficult to adhere to. For instance, some kinds of therapy include weekly visits to psychologists, which can be difficult to fit in a family schedule, and other treatments require daily exercise (executive functioning training) which can be difficult for a child to motivate herself to do (Diamond 2012).

The second problem concerning the current focus on diagnosis is related to the very fact that many mental health conditions go hand in hand with significant *stigmatization* (Pescosolido et al. 2008). This is a central concern for all kinds of mental health care systems and has a negative influence on provisions of CYHC as well. As a result not only of their behavioural problems but also of any mental health diagnosis, children might have limited schooling options, be bullied by peers and experience various forms of discrimination. Given that many developmental disorders have a familial component, this may affect not only individuals but whole families. Stigma is not a simple concept to challenge as it has many faces and levels: from stigmatization within a family, to prejudice at school or within the health care system (Chisholm et al. 2012; O'Driscoll et al. 2012), which need to be addressed in a personalised way. The fear of stigma and prejudice massively influences people's decision-making in seeking help for developmental troubles of their children (Corrigan 2004). As a result, parents seem to postpone their first visit to CYHC physicians, leading to a situation in which problems are already rather severe and prevention very often is no longer an option. Parents often have to make a difficult choice between seeking help and facing stigmatization or avoiding help and possibility of stigmatization, thereby running the risk of not solving or tackling the problems, but inducing prolonged suffering and decreased quality of life. Some parents also fear that they themselves will be stigmatised as "bad parents", blamed for a psychiatric diagnosis of their children and that potentially their child can even be taken away from them (Syurina et al. 2015).

For these reasons the CYHC, rather than focusing on diagnostic medicine, should ensure that all children who experience developmental problems receive adequate help and support. To this end, it is important to acknowledge that every family is unique and that different children with similar symptoms and similar challenges in their everyday functioning might not only have a different genetic make-up, but also grow up in different environments with different household structures, different schools and different cultures. In order to provide equal help and support, CYHC should focus on preventive, predictive and individually-based approaches that take all relevant differences in the lives of children into account (Committee on Psychosocial Aspects of Child Family Health Task Force on Mental Health 2009). Therefore, CYHC should move away from curative-oriented frameworks towards earliest possible symptom management and when possible towards prevention of conditions.

4.6.2 Limited Integration of Scientific Advances

Out of the large and increasing number of recent findings on child development in genetics and neurobiology, only very few are currently used in everyday CYHC practice. It has been shown that only around 14 % of all scientific advances and discoveries ever enter everyday practice taking an average of 17 years to reach that stage (Westfall et al. 2007). Moreover, only 5 % of all advances in basic science labelled as "highly promising and innovative" were actually licensed for use in clinical work over a period of 15 years (Contopoulos-Ioannidis et al. 2008). This general picture seems also to apply to CYHC. We believe this indicates that there is a great need for better translation and integration of basic scientific knowledge into daily practice of clinical work. In some cases, the advances could take the form of new screening techniques (applied to all children), for instance as part of post-natal genetic screening (heel-prick screening). In others, advances could involve the translation of knowledge for use in daily practice (i.e. integrating knowledge on the factors influencing development of the disease and how they can be modified). Ideally, this will allow for better estimations of the developmental pathway of individual children, make early interventions possible (e.g. suitable modifications of environmental factors) and thereby better prevent suffering.

4.6.3 Need for Personalized Healthcare

A more personalized health care can be achieved by a thorough assessment of each individual case. Indeed, personalisation of care is often mentioned in research and policy documents, but there is no general consensus on what it means as it has many different aspects. In the following paragraphs we describe our own conception of the term and its application.

Even though many different terms have been used in the literature to describe personalization, we prefer to adhere to 'personalized health care', because this is the broadest of all, uniting possible varieties of personalised services (Pokorska-Bocci et al. 2014). The general idea of personalized health care is to tailor and to adjust health services and information to the specific features and needs of each individual (based on their genetic and neurobiological profile and environment) and to thereby bring about particular health improvements. Personalized health care implies the integration of knowledge about a person's genetic baseline risks, the neurobiological processes that underlie behavioural development, the various environmental influences on a person and the complex interactions of both biological and environmental features.

We will here make some suggestions of ways in which the principles of personalised health care could be included in current CYHC practices. Currently, most attention within CYHC is paid to the assessment of the child's environment as well as her psychological condition. We believe that *integrating information about a child's genetic make-up and relevant neurobiological features* into these assessments will lead to a better and more objective evaluation of each personal case and provide room for better tailored means of prevention and intervention (Campbell and Rohrbaugh 2013). A better and broader use of the available data, such as growth and mental development, in combination with the minimization of harmful environmental exposures, are keystones for successful prevention (Vardas 2008).

Nonetheless, collecting yet more data about specific children is not enough for truly personalized health care. In addition to acquiring more information about the child, accurate *interlinks* between different data should be made and analysed holistically in order to gain full insight into personal vulnerabilities and protective factors. Both strengths and weaknesses of each child are of great importance for the development of preventive strategies: one should be aware of risks in order to minimise them, but one should also look at children's strengths in order to encourage protective mechanisms. Thus, the way data is recorded and used is of paramount importance.

Currently a lot of general data is collected within CYHC (multi-stakeholder questionnaires, regular bodily measures, etc.), but due to lack of uniformity of terminology and language, their use is sub-optimal. For instance, when reporting family situations in the files, some physicians note only the family composition, while others make remarks about the general relations and yet others emphasize the family medical history more. This does not mean that they ask different questions, but it rather reflects their perceptions of which data are valuable and which are less useful. Such discrepancies, together with the fact that each child can be assessed by several physicians at different time points, make it difficult to make longitudinal links between the various factors. *A good homogenous registration* system can help practitioners to make better use of the data in their assessments.

To sum it up, the personalized CYHC of the future should aim at providing the right intervention, to the right person at the right time.

4.7 Parents and the New CYHC

The suggestions for the modification of the CYHC system discussed in the previous paragraphs also entail changes in the position and responsibility of the parents. In order for CYHC to become more personalised, it needs to become more participatory and empowering and bring parents on board as partners. The responsibility of the parents to care for their child and to seek timely help when necessary goes hand in hand with the CYHC's responsibility to provide information about the timing of interventions, what type of help can be provided and how to choose among many options. That is why both types of responsibility, that of the parents and that of professionals, are discussed together.

To begin with, the interpretation of parental responsibility changes in the new paradigm. Parents have a responsibility to play a key role in supporting timely collection of the relevant information about their child's development as they are the ones with the best knowledge of the child. They are the ones spending the most time with the child, and thus are first to notice early signs of possible difficulties (i.e. sudden reduction of motivation, mood swings, disruptive behaviour), which they should bring to the attention of CYHC professionals. Second, parents and CYHC physicians need to form a partnership. To do so, parents have a responsibility to take a more active and informed part in providing help for their children. They should take more control over the choice of intervention and change or identify and minimize negative environmental influences as far as possible. For instance, if extreme usage of TV and computer games places the child at risk of sleep disturbance, parents should discuss intervention options with the physicians. In some cases, there can be a need to reinforce house rules to reduce such exposure. However, if certain conditions cannot be modified, parents need to contribute to choosing other possibilities (home support or school-based interventions). Increasing participation from parents (through a dialogue with CYHC physicians) in the care-linked decision-making process can help to reach better health outcomes for the child (King et al. 2014), and part of the responsibility to make this participation work lies with the parents.

At the same time, however, we need to emphasize that parents do not only carry extended responsibilities, but should be allowed and enabled to formulate and enforce their own claims on the CYHC system. In order to be able to do so, the CYHC has responsibilities as well: it should make sure that parents actually understand the educational or developmental information provided to them and have insights in the consequences of this information for the child. This holds in particular for relevant findings from contemporary genetics and neuroscience. To date, knowledge of these findings among the general public is quite low and there are many misconceptions and misinterpretations (Syurina et al. 2011). Our investigation of knowledge of and attitudes towards family medical history among parents of young children identified many inconsistencies in parents' ideas on the development and detection of conditions and disorders (Syurina et al. 2015). For example, some parents believe that some conditions like ADHD or diabetes are almost exclusively linked to environmental conditions like unhealthy eating or overconsumption of sugar. Others identified conditions such as asthma or obesity as having almost exclusively genetic bases. One respondent even believed that prenatal screening through ultrasound and blood tests could identify all genetic and developmental risks over the child's whole life, so there was no need for the CYHC system (Syurina et al. 2015). Indeed, to increase the effectiveness and efficiency of CYHC, parental health literacy must be improved. For health professionals this means that they have a responsibility to not only provide larger amounts of health information, but also to ensure that it is individually adapted to the background knowledge of individual parents, and to give them greater ability to ask questions. In a recent review assessing the effectiveness and efficiency of community based CYHC, researchers identified several important implications and directions for change, such as the increased involvement of various stakeholders (especially parents) in care provisioning and health management (Chisholm et al. 2012). The CYHC needs to assist in the development of a closer relationship between parents and care providers by investing more time in contact that is truly productive, in providing information and in finding other ways to challenge negative parental beliefs about the system (i.e. rapidly addressing concerns). This in turn will lead to increased levels of trust and better outcomes for the child, as parents would be more inclined to search advice earlier and have better adherence to the health advice.

4.8 Conclusions

Children can benefit a lot from timely preventive interventions. The CYHC aims to monitor the development of the child and to detect warning signs in a timely manner in order to implement these interventions. The present diagnosis-oriented approach of CYHC systems in the Western world in general, and in the Netherlands in particular, is not ideal and should be adapted so that it can make use of the new predictive and preventive knowledge provided by advances in genetics and neuroscience. Within the current system children experiencing problems need to fit within the constraints of an official DSM diagnosis before they can access mental health care, which is at the moment primarily curative in nature rather than preventive. Still, many children with psycho-social difficulties in their everyday life (at home and/or at school) have symptoms that impede their development, though these symptoms may not yet cluster to any DSM-diagnosis. For them, currently there is little help available, and their families are left to struggle on their own. Moreover, such an approach leaves little to no space for prevention of problems because it focuses on treatment of conditions.

Another area for improvement is the level of involvement of parents in care provision. At the moment this involvement is rather low: there is little room for active participation and health professionals hardly encourage it. This leads to a situation where parents are given little information about their child's health status and potential risks, cannot always voice their concerns and do not have an opportunity to meaningfully participate in the decision-making process. All of this in turn can contribute to misconceptions of the CYHC system, to mental health stigma and to a corresponding reluctance to search for help. As a consequence, the opportunities for prevention are limited, causing additional suffering to the child and family. To improve this situation, there is a need to invest in establishing more trusting and safe relationships between CYHC physicians and the parents as well as in empowering family members to participate in public health care provision for children.

A move away from a diagnosis-oriented approach with a low level of parental involvement towards a more predictive and personalised health care provision for children is necessary. In order for that to be possible, several important steps need to be undertaken. Among them are: further introduction of biology-based assessments, holistic integration and use of varied sources of data and a broader and deeper use of the data. Last but not least, health professionals must establish a closer relationship with other relevant stakeholders who are involved in the upbringing of the child. All these factors combined should lead to better predictive and preventive strategies in child and youth health care.

Author Contributor Statement The initial concept for the article was developed by Frans J.M. Feron. The first draft was written by Elena V. Syurina and then was reviewed by both authors. Frans J.M. Feron brought in additional knowledge of the current situation within child and youth health care (CYHC) and Elena V. Syurina was responsible for integrating the neuroscience and genetics knowledge as well as the new concept of CYHC. Both authors contributed to the conclusions and recommendations part of the chapter.

References

Achenbach, T.M., and C. Edelbrock. 1981. *Child behavior checklist*. Burlington: University of Vermont.

Allis, C.D., T. Jenuwein, D. Reinberg, and M.-L. Caparros. 2007. *Epigenetics, cold spring harbor*. New York: Laboratories Press.

American Psychiatric Association, and DSM 5 Task Force. 2013. *Diagnostic and statistical manual of mental disorders: DSM-5*. From http://dsm.psychiatryonline.org/book.aspx?bookid=556.

Andersen, S.L. 2003. Trajectories of brain development: Point of vulnerability or window of opportunity? *Neuroscience & Biobehavioral Reviews* 27(1): 3–18.

Archer, T., M. Oscar-Berman, and K. Blum. 2011. Epigenetics in developmental disorder: ADHD and endophenotypes. *Journal of genetic syndrome & gene therapy* 2(104): 1–17.

Bayley, N. 1993. *Bayley scales of infant development manual: Psychological corporation*. San Antonio: Harcourt Brace & Co.

Bee, H.L., and D.R. Boyd. 2000. *The developing child*. Boston: Allyn and Bacon.

Ben-Arieh, A., N.H. Kaufman, A.B. Andrews, R.M. George, B.J. Lee, and L.J. Aber. 2013. *Measuring and monitoring children's wellbeing*, vol. 7. Berlin: Springer Science & Business Media.

Birch, A.M., N.B. McGarry, and Á.M. Kelly. 2013. Short-term environmental enrichment, in the absence of exercise, improves memory, and increases NGF concentration, early neuronal survival, and synaptogenesis in the dentate gyrus in a time-dependent manner. *Hippocampus* 23(6): 437–450.

Boomsma, S., and F. Pijpers. 2008. Een visie op de jeugdgezondheidszorg als publieke zorg voor de jeugd. *Tijdschrift voor Gezondheidswetenschappen* 86(7): 377–380.

Campbell, W.H., and R.M. Rohrbaugh. 2013. *The biopsychosocial formulation manual: A guide for mental health professionals*. New York: Routledge.

Chiang, M.-C., K.L. McMahon, G.I. de Zubicaray, N.G. Martin, I. Hickie, A.W. Toga, and P.M. Thompson. 2011. Genetics of white matter development: A DTI study of 705 twins and their siblings aged 12 to 29. *NeuroImage* 54(3): 2308–2317.

Chisholm, K.E., P. Patterson, C. Torgerson, E. Turner, and M. Birchwood. 2012. A randomised controlled feasibility trial for an educational school-based mental health intervention: study protocol. *BMC Psychiatry* 12(1): 23.

Coker, T.R., T. Thomas, and P.J. Chung. 2013. Does well-child care have a future in pediatrics? *Pediatrics* 131(Supplement 2): S149–S159.

Committee on Psychosocial Aspects of Child Family Health Task Force on Mental Health. 2009. Policy statement—the future of pediatrics: Mental health competencies for pediatric primary care. *Pediatrics* 124: 410–421.

Contopoulos-Ioannidis, D.G., G.A. Alexiou, T.C. Gouvias, and J.P.A. Ioannidis. 2008. MEDICINE: Life cycle of translational research for medical interventions. *Science* 321: 1298–1299.

Corrigan, P. 2004. How stigma interferes with mental health care. *American Psychologist* 59(7): 614.

Crews, F., J. He, and C. Hodge. 2007. Adolescent cortical development: A critical period of vulnerability for addiction. *Pharmacology Biochemistry and Behavior* 86(2): 189–199.

Cutler, D., and E. Meara. 2001. *Changes in the age distribution of mortality over the 20th century*. Cambridge: National Bureau of Economic Research.

Cutler, D., and G. Miller. 2005. The role of public health improvements in health advances: The twentieth-century United States. *Demography* 42(1): 1–22.

Davidson, E.H. 2012. *Gene activity in early development*. San Diego: Elsevier.

Diamond, A. 2012. Activities and programs that improve children's executive functions. *Current Directions in Psychological Science* 21(5): 335–341.

Dute, J. 2008. De wet publieke gezondheid. *Tijdschrift voor Gezondheidsrecht* 32: 576–591.

Folio, M.R., and R.R. Fewell. 1974. *Peabody developmental motor scales*. Nashville: George Peabody College for Teachers.

Fridlizius, G. 1989. The deformation of cohorts: nineteenth century mortality decline in a generational perspective. *Scandinavian Economic History Review* 37(3): 3–17.

Goodman, R. 2001. Psychometric properties of the strengths and difficulties questionnaire. *Journal of the American Academy of Child & Adolescent Psychiatry* 40(11): 1337–1345.

Goodman, A., R. Joyce, and J.P. Smith. 2011. The long shadow cast by childhood physical and mental problems on adult life. *Proceedings of the National Academy of Sciences* 108(15): 6032–6037.

Green, J., T. Charman, A. Pickles, M.W. Wan, M. Elsabbagh, V. Slonims, and R. Booth. 2015. Parent-mediated intervention versus no intervention for infants at high risk of autism: A parallel, single-blind, randomised trial. *The Lancet Psychiatry* 2(2): 133–140.

Greenberg, M.T., R.P. Weissberg, M.U. O'Brien, J.E. Zins, L. Fredericks, H. Resnik, and M.J. Elias. 2003. Enhancing school-based prevention and youth development through coordinated social, emotional, and academic learning. *American Psychologist* 58(6–7): 466.

Hagan, J.F., J.S. Shaw, and P.M. Duncan. 2008. *Bright futures: Guidelines for health supervision of infants, children, and adolescents*. Elk Grove Village: American Academy of Pediatrics.

Hanson, J.L., N. Adluru, M.K. Chung, A.L. Alexander, R.J. Davidson, and S.D. Pollak. 2013. Early neglect is associated with alterations in white matter integrity and cognitive functioning. *Child Development* 84(5): 1566–1578.

Illingworth, R.S. 2013. *The development of the infant and the young child: Normal and abnormal*. London: Elsevier Health Sciences.

Kail, R.V. 2001. *Children and their development*. Upper Saddle River: Prentice Hall.

Kieling, C., H. Baker-Henningham, M. Belfer, G. Conti, I. Ertem, O. Omigbodun, and A. Rahman. 2011. Child and adolescent mental health worldwide: Evidence for action. *The Lancet* 378(9801): 1515–1525.

King, G., M. Currie, and P. Petersen. 2014. Child and parent engagement in the mental health intervention process: A motivational framework. *Child and Adolescent Mental Health* 19(1): 2–8.

Knodel, J.E. 2002. *Demographic behavior in the past: A study of fourteen German village populations in the eighteenth and nineteenth centuries, vol. 6*. Cambridge: Cambridge University Press.

Knudsen, E. 2004. Sensitive periods in the development of the brain and behavior. *Journal of Cognitive Neuroscience* 16(8): 1412–1425.

Martel, M., M. Nikolas, K. Jernigan, K. Friderici, I. Waldman, and J. Nigg. 2011. The Dopamine Receptor D4 gene (DRD4) moderates family environmental effects on ADHD. *Journal of Abnormal Child Psychology* 39(1): 1–10.

Mason, S., & F.C. Zhou. 2015. Genetics and epigenetics of fetal alcohol spectrum disorders. *Frontiers in Genetics* 6:146.

McLeod, S. 2007. Skinner: Operant conditioning. *Simply Psychology*.

Norman, R.E., M. Byambaa, R. De, A. Butchart, J. Scott, and T. Vos. 2012. The long-term health consequences of child physical abuse, emotional abuse, and neglect: A systematic review and meta-analysis. *PLoS Medicine* 9(11): e1001349.

O'Driscoll, C., C. Heary, E. Hennessy, and L. McKeague. 2012. Explicit and implicit stigma towards peers with mental health problems in childhood and adolescence. *Journal of Child Psychology and Psychiatry* 53(10): 1054–1062.

Paul, L.K., W.S. Brown, R. Adolphs, J.M. Tyszka, L.J. Richards, P. Mukherjee, and E.H. Sherr. 2007. Agenesis of the corpus callosum: Genetic, developmental and functional aspects of connectivity. *Nature Review Neuroscience* 8(4): 287–299.

Pescosolido, B.A., P.S. Jensen, J.K. Martin, B.L. Perry, S. Olafsdottir, and D. Fettes. 2008. Public knowledge and assessment of child mental health problems: Findings from the national stigma study-children. *Journal of the American Academy of Child & Adolescent Psychiatry* 47(3): 339–349.

Piaget, J., and M.T. Cook. 1952. *The origins of intelligence in children*. New York: International Universities Press.

Pokorska-Bocci, A., A. Stewart, G.S. Sagoo, A. Hall, M. Kroese, and H. Burton. 2014. 'Personalized medicine': what's in a name? *Personalized Medicine* 11(2): 197–210.

Raznahan, A., D. Greenstein, N.R. Lee, L.S. Clasen, and J.N. Giedd. 2012. Prenatal growth in humans and postnatal brain maturation into late adolescence. *Proceedings of the National Academy of Sciences* 109(28): 11366–11371.

Roth, T.L., F.D. Lubin, A.J. Funk, and J.D. Sweatt. 2009. Lasting epigenetic influence of early-life adversity on the *BDNF* gene. *Biological Psychiatry* 65(9): 760–769.

Sawicki, M.P., G. Samara, M. Hurwitz, and E. Passaro. 1993. Human genome project. *The American Journal of Surgery* 165(2): 258–264.

Shonkoff, J.P., and S.J. Meisels. 2000. *Handbook of early childhood intervention*. New York: Cambridge University Press.

Smidt, S. 2013. *The developing child in the 21st century: A global perspective on child development*. London/New York: Routledge.

Snyderman, R., and J. Langheier. 2006. Prospective health care: The second transformation of medicine. *Genome Biology* 7(2): 104.

Steinberg, L. 2007. Risk taking in adolescence new perspectives from brain and behavioral science. *Current Directions in Psychological Science* 16(2): 55–59.

Syurina, E., I. Brankovic, N. Probst-Hensch, and A. Brand. 2011. Genome-based health literacy: A new challenge for public health genomics. *Public Health Genomics* 14(4–5): 201–210.

Syurina, E., K. Hens, and F. Feron. 2013. Literature review: Use of family history for primary paediatric care as the next step towards use of genomics in healthcare. *Current Pediatric Reviews* 9(4): 353–372.

Syurina E.V., A. Gerritsen, K. Hens, W. Dondorp, and F.J.M. Feron. 2015. What about FH of my child? Parents' opinion on family history collection in preventive primary pediatric care. *Personalized Medicine* 12(4): 327–337.

Twardosz, S., and J.R. Lutzker. 2010. Child maltreatment and the developing brain: A review of neuroscience perspectives. *Aggression and Violent Behavior* 15(1): 59–68.

van Mil, N.H., R.P.M. Steegers-Theunissen, M.I. Bouwland-Both, M.M.P.J. Verbiest, J. Rijlaarsdam, A. Hofman, and H. Tiemeier. 2014. DNA methylation profiles at birth and child ADHD symptoms. *Journal of Psychiatric Research* 49: 51–59.

Vardas, P.E. 2008. Personalised health care: an emerging aim. *Hellenic Journal of Cardiology* 49(6): 453.

Westfall, J.M., J. Mold, and L. Fagnan. 2007. Practice-based research–"Blue Highways" on the NIH Roadmap. *JAMA* 297: 403–406.

Youngson, N.A., and E. Whitelaw. 2008. Transgenerational epigenetic effects. *Annual Review of Genomics and Human Genetics* 9: 233–257.

Chapter 5
Raising Self-Controlled Children. A Philosophical Analysis of Neuroscience and Social Psychology Perspectives

Dorothee Horstkötter

Abstract In recent decades, self-control has received increasing attention as it can safeguard child-development and human wellbeing. Researchers from various disciplines – philosophy, neuroscience and social psychology – have investigated what self-control is, how it is generated and how it can be stimulated. This chapter critically reflects on recent discussions of the concept and the science of self-control while investigating their bearings on the question of whether parents have a responsibility to raise self-controlled children and what that would entail. The argument put forth is that current social psychology and neuroscience largely investigates *controlled* behaviour but ignores the prefix *self*. Consequently, a more comprehensive understanding of the term that does justice to both aspects is provided. This gives rise to two different sets of educational goals. Firstly, raising self-controlled children entails teaching them strategies to overcome temptation. Secondly, it requires that parents support children to develop a self that sets its own goals, reflects on these goals and considers them as reasons for action.

Keywords Agency • Akrasia • Autonomy • Combat-model • Compulsion • Constitution-model • Delay of gratification • Education • Inhibition • Korsgaard, Ch • Marshmallow test • Motivation • Neuroscience • Philosophy of action • Prefrontal cortex • Self • Self-control • Self-determination • Self-control strategy • Social psychology • Recklessness • Velleman, D • Weakness of will

D. Horstkötter (✉)
Department of Health, Ethics and Society, School of Mental Health and Neuroscience (MHeNS), University of Maastricht, Postbox 616, 6200 MD, Maastricht, The Netherlands
e-mail: d.horstkoetter@maastrichtuniversity.nl

K. Hens et al. (eds.), *Parental Responsibility in the Context of Neuroscience and Genetics*, International Library of Ethics, Law, and the New Medicine 69,
DOI 10.1007/978-3-319-42834-5_5

5.1 Introduction

In the last decades, self-control has received increasing attention as an important means to safeguard human wellbeing. In other words, lack of self-control has been identified as the cause of much personal and social misery (Baumeister et al. 1994). Moreover, childhood self-control is frequently considered a good predictor of the self-control that people exhibit throughout their lives (Casey et al. 2011; Moffitt et al. 2011). This raises the question of whether parents have a responsibility to raise self-controlled children, that is, to instill self-control in their children. If they do have such a responsibility, what does it entail? Researchers from a variety of disciplinary backgrounds – philosophy, neuroscience and social psychology – have discussed what self-control is, how it is generated, and how it can be stimulated (e.g., Aron et al. 2004; Baumeister and Heatherton 1996; Hare et al. 2009; Henden 2008; Mele 1995). At first glance, there seems to be widespread agreement regarding the meaning and the relevance of self-control. It has frequently been linked to will-power and has been taken to refer to the successful overriding of inadequate spontaneous responses, for the sake of desirable yet distant goals. At the same time, however, it has also been argued that self-control is a complex concept denoting not only processes of successful goal-achievement, but also the development of understanding of what counts as a desirable goal and what it means when behaviour is controlled by someone, that is, by a self (Cervone et al. 2006; Deci and Ryan 2000; Velleman 2000). In that sense, self-control denotes control *of* the self *by* the self (Horstkötter 2015) referring both to the object and the subject of human behaviour. As a consequence, a conceptualization of self-control that is able to incorporate various tiers and which differentiates between relevant aspects of what can reasonably be called self-controlled behaviour, may best encapsulate the full meaning of the term (Kennett 2001).

In this chapter, I will critically analyse recent views of the concept of self-control and the science behind it and I will examine the bearings that current conceptual thought and research findings have on the question of whether parents have a responsibility to raise self-controlled children. I will also look at possible implications and the effects of fostering self-control in children and to support its development in young people. To this end, a series of studies are presented that emphasize the impact of childhood self-control on the future wellbeing of adults in various areas of life and that highlight the need for the active development of self-control in children from an early age. I will then provide a brief overview of up-to-date research data in neuroscience and social psychology regarding the neural basis and the psychological and environmental conditions of self-controlled behaviour. These scientific approaches towards self-control will be critically discussed while emphasizing crucial conceptual shortcomings. It is particularly interesting that these approaches examine 'controlled' behaviour while the prefix 'self' is hardly given any explicit attention. Therefore, a more comprehensive understanding of the term is developed which can do justice to the self being simultaneously the object and the subject of self-controlled behaviour. To that end, I will develop two lines of argument. First, I will sketch a three-tiered picture of self-control that covers weakness of will, compulsion and recklessness as distinct varieties of self-control failure. Recklessness is of particular importance in this regard because it denotes a failure of goal-setting or goal-evaluation rather than goal-achievement. Reversely, this indicates that successful self-control is also about goal-setting and evaluation, rendering the self the indispensable evaluating subject of self-controlled behaviour. Second, I will argue that in order to fully understand self-control, it is important to acknowledge what it means for a person to be an agent. To that end, the distinction between action and event will be introduced and briefly elaborated on.

Equipped with this philosophically informed background, I will then revisit recent neuroscience and social psychology research and critically ascertain their alleged implications for a parental responsibility of raising self-controlled children. I will point out that different conceptualizations of self-control account for and require different sets of educational goals. Recent findings in neuroscience and social psychology can indeed have repercussions for successful goal-achievement in situations of temptation. However, this does not hold when self-control refers to the reasons people have for their behaviour and to their evaluation of what is a desirable goal for them. Here, neuroscience and social psychology findings have no direct bearing.

5.2 On the Importance of Childhood Self-Control

For several decades and with increasing frequency, not only scientific, but also popular media and popular science literature have made the public aware that self-control is of paramount importance for achieving wellbeing and success in life in a variety of areas (Baumeister and Tierney 2011; McGonigal 2012). Similarly, lack of self-control has been deemed to be the main cause of many personal and social problems of our time, including health problems, economic miseries and even criminal behaviour (Baumeister et al. 1994; Gottfredson and Hirschi 1990). Apparently, this does not only hold true for adults, but better self-control has been linked to greater agreeableness, conscientiousness and open-mindedness also in children. In line with this view, it has been shown that people who as children had comparably good self-control, have better lives as adults.

To this avail, longitudinal studies spanning over 15–30 years of the development of individual participants have been conducted to compare adult functioning with preschool self-control (Loeber et al. 2012; Moffitt et al. 2011; Slutske et al. 2012). Preschool self-control has been defined largely by measures of children's impulsivity, as well as, their emotional liability, restlessness and attention span. When comparing the behavioural styles of young children with their lives and behaviour in adult years, it appeared that those who had more self-control decades before, that is, those who were less impulsive, more emotionally stable and who had longer attention spans, have more success in adult life. They have better physical health, are wealthier and display on average less incidence of, e.g., substance dependence, gambling or criminal behaviour (Loeber et al. 2012; Moffitt et al. 2011; Slutske et al. 2012).

Studies that investigated children's self-control in terms of their ability to delay gratification (Mischel et al. 1989) report similar effects. Individual differences on self-control could not only be determined already at kindergarten-age, but also appeared to correspond to differences in cognitive and social performances many years later (Mischel et al. 1996). More recently, differences in brain function between low and high delayers have been highlighted. That is, those who as children differed in their ability to delay gratification showed corresponding differences in their behavioural patterns many years later; their brains could be shown to exhibit corresponding functional differences (Casey et al. 2011).

Despite these correlational relationships and the idea that childhood self-control is a good predictor of adult outcomes, self-control has also been shown to be malleable and to change over time. It is not predetermined. It develops via natural maturation and it can improve through specially designed brain training programmes (Berkman et al. 2012). Against this double background, a pivotal conclusion of these studies points towards the importance of early and effective intervention in children's capacity to exercise self-control. Enhancement of self-control should be an important target of both childhood and adolescence interventions, according to the forceful advocacy of the authors of one of the longitudinal studies (Moffitt et al. 2011). Most recently, such scientific findings are readily received by the public and the media. More in particular, parenting guides are published not only to inform parents and professionals about the importance of childhood self-control, but also to provide instructions on how to teach it to children (e.g. Aha! Parenting 2013; Browers 2012). Before critically discussing this tendency and the apparent urgent need to teach children self-control, I will analyse what is meant by self-control and which types of behaviours are perceived as consistent with the exercise of self-control.

5.3 Neuroscience and Self-Control

In the last 10 years, neuroscience research on self-control has intensified and has received more and more public attention. Increasingly, claims are being made regarding its implications for the set-up of behavioural interventions and the design of policy measures. This raises the question whether, and if so how, parents have a responsibility to aim at the development of self-control in their children by taking corresponding neuroscience findings on children's self-control into account.

The main question of neuroscience research regards the location of self-control in the brain and the activity of these areas under different conditions. That is, neuroscientists try to identify which brain regions are involved whenever people can be said either to exercise or to fail to exercise self-control. Self-controlled behaviour can then be explained by the pattern of activation of relevant brain regions. To this end, neuroscience research focuses on inhibitory control as an important subcategory of self-control (Berkman et al. 2012) and on that basis defines whether any behaviour ought to count as either an instance of self-control or of self-control failure. According to current definitions, inhibition denotes the cancelling of intended movements, the suppression of inappropriate responses (Aron et al. 2004), or the ability to override a dominant response in order to enact a subdominant one (Berkman et al. 2012).

Self-control has further been connected to cognitive control as a crucial facilitator of delayed gratification. According to this understanding, it denotes "the ability to suppress competing inappropriate thoughts or actions in favour of appropriate ones" (Casey et al. 2011: 14998). These abilities have been mainly investigated by lesion and brain-imaging studies. In carefully designed experimental settings, participants are given so-called self-control tasks whereupon their brain-activity is measured, or – in case of lesion studies – their performances are compared to their known brain deficits. Typical self-control tasks in neuroscience include the so-called *Go/No-go* and *stop signal* tasks. Here, participants are presented with rapid sequences of predefined cues and are asked to respond speedily to them. They are shown, for example, a series of letters and are asked to respond [Go], e.g. press a button, to some of these cues such as the letters A,B, or C and to inhibit the response [No-go], that is not press the button, in case some other letters are presented. Alternatively, they may be asked to stop responding whenever a particular signal, for example a sound, is given [stop-signal]. How well people are able to inhibit, i.e. exercise self-control, is then measured as a function of either the errors they make on the no-go trial (how often they press the button when they are supposed not to do so), or of the time needed to react to the stop-signal.

Performance of such tasks while being brain scanned delivers information about brain regions involved, and the brain activity correlated with the performance. Apparent differences in the neurobiological substrate are in turn described as underlying the observable differences between people's behaviour and their different success rates in self-controlled behaviour. Most prominently, various regions in the prefrontal cortex (PFC), typically connected to cognitive processes, have been identified as the main locations of self-controlled behaviour (Aron et al. 2004). Hare et al. (2009) specified these findings by pointing out that one PFC sub-region (ventromedial) is considered to process the incoming signal on the short-term while another sub-region (dorsolateral) is supposed to incorporate long-term considerations and on that basis modulate the incoming signal. Such a system may allow for the explanation of differences between self-controllers and non-self-controllers on the neural level in two ways: either the activity within the highlighted brain regions differs between participants or there are differences in the interconnections between the sub-regions involved.

In addition to the involvement of prefrontal regions, Casey et al (2011) point towards the relevance of the ventral striatum which is a deep brain system. This is considered relevant for the processing of emotional and rewarding cues. Amplified activity in this emotional region may reflect a high sensitivity to emotional and rewarding cues and in turn negatively influence activity at the prefrontal level and people's ability to suppress emotionally loaded thoughts and actions. At the neural level, differences in behavioural self-control are then to be explained by interpersonal differences between the relative activity of the prefrontal and of the deep brain system. Diminished prefrontal activity in conjunction with amplified activity of the deep brain system may then be the neural basis of any uncontrolled behaviour, that is, behaviour that follows immediate emotional triggers yet fails to display cognitive control. These differences in self-control related brain-activity can be applied in two ways (Berkman et al. 2012). First, they can inform the development of specialized training programmes that specifically target the brain areas deemed relevant. Second, childhood self-control and related brain activity can be seen as a predictor of adult self-control and help determine who is, and who is not, in need of behavioural interventions.

5.4 Social Psychology and Self-Control

While neuroscientists are looking for neural correlates and neural underpinnings of self-controlled behaviour, social psychologists are mainly interested in the features that constitute self-controlled behaviour, the conditions that may facilitate or hinder it and the actions that people do when they behave in a self-controlled way. In the following I will briefly present the main approaches and findings in social psychology in this area and reflect on their potential implications for early self-control improvement.

Research on behavioural self-control has been on the agenda of social psychology for the last 30 years. Self-control, as its early modern pioneers argued, is indispensable for the understanding of human behaviour (Carver and Scheier 1998). Essentially, it has been considered to function like a thermostat. By way of an informational feedback loop people's actual states (behaviour) are compared to some target states (desired goal). In case of a discrepancy people adapt their behaviour so that they are still able to achieve the respective goal. This basic picture of action control turned out to be highly influential in subsequent self-control research. In their review of self-regulation failure, Baumeister and colleagues considered self-regulation to be crucial for goal-achievement and to entail the alteration of people's spontaneous responses (Baumeister et al. 1994). To that end, they distinguished several phases. For instance, people monitor their actual response, they keep in mind certain standards or goals that are to be achieved. Finally, they adapt or control their response so that potential divergences are diminished and the goal is achieved.

Numerous experiments have since then been conducted to investigate the empirical basis of this understanding of human behaviour and the conditions under which self-control is successful or failing. These have led to the so-called *strength model of self-control*. According to this model, people cannot only bring about changes in their current behaviour and reorient in case their actual doings keep them from achieving their goals; but sequential, yet unrelated acts of self-control diminish their strength to perform equally well in later self-control tasks. For example, people who in experimental set-ups were first asked not to think of a white bear (thought-control) were afterwards less able to suppress their emotional response to a disgusting movie (emotion-control) or to squeeze a handgrip and overcome physical discomfort (Muraven et al. 1998) than people who were not given the initial thought-control task and who were therefore not depleted. As a consequence, self-control has been considered a matter of strength and self-control failure has been explained as the result of some mental weakness or a weakness of will. Often called willpower, this mental strength is thought to work like a mental muscle that becomes depleted after exertion, replenishes after rest, and can be trained to become stronger in the long run (Baumeister 2012; Hagger et al. 2010; Muraven and Baumeister 2000).

Related research that builds on the early work of Mischel and colleagues (e.g. Mischel et al. 1972) focused on the concrete actions that people, particularly young children, perform when they exercise self-control, and questioned whether some kinds of action may better support distant goal-achievement than others. Their research uncovered a variety of specific strategies that one may invoke in order to delay gratification and achieve later but greater rewards. The test has become famous as "The Stanford Marshmallow Experiment". It investigated, for example, the degree to which young children were able to withhold from taking a tasty reward (such as a marshmallow) while left unsupervised, after being told that leaving the reward untouched would lead to them being offered two rewards. Thanks to this experiment it was discovered that children are better at waiting when they, for example, cognitively refigured a marshmallow as a white and puffy yet inconsumable cloud, moon or ball, or when they thought of something pleasurable instead (Mischel et al. 1989, 2014). Teaching children such strategies improved their ability to wait and made them significantly better self-controllers than their uninstructed peers.

5.5 An Intermediate Conclusion ....

These findings in social psychology preceded and motivated research in neuroscience such as the previously presented. Neuroscientists interested in the neural underpinnings of self-control refer back to both the strength-model (e.g. Berkman et al. 2012) and the delay of gratification paradigm (e.g. Casey et al. 2011). Therefore, from an interdisciplinary perspective there seems to be widespread agreement regarding the relevance and the meaning of self-control. Furthermore, the implications of how to improve self-controlled behaviour appear to be straightforward and shared among the two disciplines. On the public policy level, it seems advisable to arrange default settings so that people are less in need of self-control which will make self-controlled decisions much easier than not self-controlled decisions (Moffitt et al. 2011). This may avoid the depletion of self-control strength or prevent the activation of emotional brain areas. The banning of smoking in public areas or of sweets from schools are examples of this. In addition, in order to achieve comparable outcomes, for example to make people quit smoking or reduce their sweets consumption, it seems conducive to teach people distinct self-control strategies (Strayhorn 2002), i.e. train their brains according to insights on the neural generation of self-control (Berkman et al. 2012), or have them practice and strengthen the self-control 'muscle' (Baumeister 2012; Muraven and Baumeister 2000).

In accordance with this perspective, a particular focus should be on children and the implementation of early intervention programmes. Children are more amendable to interventions and their brains are much more plastic than those of adults, so the argument goes (Berkman et al. 2012; Moffitt et al. 2011). In addition, suggestions have been made that these findings may be used for the set-up of interventions for children with self-control related psychopathologies. In their case, fMRI (functional magnetic resonance imaging) scans could be used to test individual brain maturity and intervention readiness before any training is offered, as well as, to trace potential increases in self-control at the neural level during or after the implementation of self-control interventions (Berkman et al. 2012).

The scientific evidence supporting a belief in people's ability to prevent and suppress impulses or emotions and the possibility to acquire such capacities already during childhood appears overwhelming. Translated into the terminology of educational goals, what parents should aim for when raising their children to be self-controlled is to be able to obey distant goals, to live up to predefined rules and to suppress immediate impulses. Parents may achieve this by teaching their children self-control strategies, by confronting them regularly with self-control tasks in order to train their 'self-control muscle', or they can enrol them in self-control testing and intervention programmes, neuroscientifically informed or otherwise.

5.6 …. And Beyond

A critical look back at the original definition of self-control as employed in the scientific work discussed calls attention to further important issues that should be addressed. The scientific experiments and the resulting evidence all focus on goal-achievement and the conditions conducive to that end. However, the original definition of self-control that underlies these experiments is much broader. It covers not only goal-achievement and the corresponding suppression of emotions or alteration of actual behaviours, but it entails also an essentially normative component. Self-control has been defined as the ability to suppress *inappropriate* thoughts or responses (Aron et al. 2004; Casey et al. 2011) or as the capacity to override and prevent *undesired* thoughts, behaviours and emotions (Baumeister 2012, 1994) and to achieve *desirable* goals. These specifications of the nature of the thoughts to be suppressed, the behaviours to be overridden (inappropriate, undesired) and the goals to be achieved can inspire important subsequent questions: Which thoughts are inappropriate and what makes some behaviours or emotions undesirable? Who determines what is inappropriate and undesirable or appropriate and desirable?

These questions have not been addressed in the research of either neuroscience or social psychology. The goals have been predefined and the tasks of participants had been mainly to achieve what they were told, not to evaluate the goals or to reflect on their appropriateness or desirability. Participants have to 'go' or to 'no go' depending on the instructions. Similarly, they have to avoid thinking of white bears, not loosen the handgrip, or whatever else the respective task required them to do or leave undone. However, one may wonder why the resulting behaviours, even if successful, should count as instances of self-controlled behaviour at all. Obviously, these behaviours entail important elements of control, yet why should one say that they are paradigmatic examples of *self*-control?

When people describe themselves as being self-controlled, they typically use expressions like: *I control myself*, or *I control what I do*. Thereby, they simultaneously refer to themselves as the subject ['I'] *and* the object ['me'/'myself'] of control. They determine what goals they consider desirable or appropriate and worthwhile to achieve and so act on those determinations. In an early critique of the feedback-loop model of behavioural self-control, psychologists Ryan and Deci emphasized that in these theories, the prefix *self* has been put with a hyphen in front of the word *regulation*, however, without adding any meaning to that term (Ryan and Deci 1999). In theories of human behaviour based on the feedback-model, it is evaluated whether the person can display control, however, it is not investigated also whether she can display self-control, that is, control directed by the self rather than by others (e.g. the researchers). This is because in these theories a certain behaviour counts as a successful instance of self-control as soon as the person has achieved the goal. Yet, this is evaluated independently of whether the person considers the goal appropriate or desirable, or whether she has given it any thought at all.

Ryan and Deci try to make good for this by drawing a difference between what they term 'controlled' versus 'self-determined' behaviour (Deci and Ryan 2000, 2002). Controlled behaviour occurs due to some intra-psychic or environmental forces; it is heteronomously initiated and does not represent any true choice of the person in question. By contrast, self-determined behaviours are initiated and regulated through autonomous choices and as such are expressions of a person's values and of oneself as an agent.

The term *autonomy* has been much discussed in various philosophical disciplines. In their work, Ryan and Deci seem to rely on the Frankfurtian understanding of the term (Frankfurt 1988). According to Frankfurt, a person acts autonomously when her second order desires determine her course of action. Second order desires are those that express people's evaluation of their immediate first order desires. That is, second order desires express the decision of whether, or not, people desire to have their first order desires. Whenever self-determined, autonomous agents also act in goal-directed ways, but they do so because the achievement of the goal is important to them. It is even argued that self-determined people would experience fewer difficulties in achieving *their* goals, showed greater behavioural persistence and displayed more effective performances (Deci and Ryan 2000).

These discussions and developments in social psychology, however, can place the meaning of parental responsibility in a different perspective. If it is the subject, that is, the self of self-control that is concerned, then teaching self-control strategies, strengthening the self-control-'muscle' or training brain areas in an attempt to influence activation patterns become less relevant to the development of self-control. They may even be the wrong things to do. In this way, parents may render their children able to achieve pre-set goals. However, in order to safeguard that the self is in place, a different set of educational goals seems required instead. It seems more adequate that parents aim at their children becoming autonomous people (Betzler 2015) who are able to reflect on the goals presented to them, as well as, on their own emotional or immediate responses. Are the goals reasonable and worthwhile and should they change their behaviour in order to achieve them? Or is it the reverse: are the goals too difficult, of minor importance or maybe even undesirable for them and should they follow their immediate emotional response and adapt their long-term goals? Some illustrative examples will clarify this. Children would, for example, display more self-control if they played soccer or tennis every week because they see themselves as soccer or tennis players and develop an identity around these activities, than children who play these sports to please their parents. In order to be successful, the latter group of children may engage some self-control technique by which they can make themselves participate in the trainings. The former group, however, is not even in need of any such techniques or skills, but their behaviour can be said to be controlled by their own conception of their self, that is by whom they are or who they want to become.

A similar analysis holds true for children who in a fit of rage shout out that they know they should count to ten – a strategy frequently taught to children to calm down – but that they will not do so because they want to be angry now and consider their anger appropriate. These children may be more *self*-controlled than those who immediately, and successfully, adopt the mentioned strategy. In the first case, the children evaluate their behaviour and the goals they have (calm down); while in the latter they do not do so. The children in the first case disagree with the goals that adults may provide them with: but disagreement is something other than self-control failure. Parents' responsibility to raise self-controlled children in this sense rather implies the importance of raising a 'self' or an autonomous person who is able to form reasons for action and the ability to act on these reasons – and maybe later change previous reasons for better reasons — rather than on installing a mere capacity of control or to develop some self-control skills or strengths.

5.7 A Brief Philosophy of Self-Control

A thorough analysis of the concept of self-control can elucidate its meaning and why it is important to acknowledge that self-control is more and something else than mere goal-achievement (Horstkötter 2015). If we want to understand the content of parental responsibility to raise self-controlled children, it is important to consider the following two aspects. First, the expression 'lack of self-control' reasonably denotes a three-partite understanding of agential failure (Kennett 2001). This idea, in turn, gives rise to the argument that successful self-control likewise encompasses three meanings of agency, that is, i) people's capacity of intentional action, ii) their capacity of goal-achievement over time, as well as iii) their well-considered reasons for action.

Second, the concept of agency is of particular importance in that it is relevantly different from mere behavioural events. An account of full-fledged agency, however, will put the idea of what is entailed in behavioural change or alteration in a different perspective than the one currently underlying the work of social psychologists and neuroscientists.

5.7.1 A Three-Partite Picture of Self-Control and Self-Control Failure

In addition to, and long before neuroscientists and social psychologists investigated self-controlled human behaviour, philosophers have been engaging with the concept and discussing its meaning. In the Nicomachean Ethics (trans. 1980: 1150a 11–13), Aristotle writes that people are self-controlled when they overcome temptations, whereas those who give in suffer from akrasia, that is, weakness of will. In a similar vein, Plato (trans. 1952, 246a, 253cff.) offered the metaphor of a charioteer who steers two horses: one obeys the voice of the charioteer smoothly and immediately, while the other one is constantly unruly and recalcitrant. The chariot is controlled whenever and as long as the charioteer and the obedient horse, with their united effort, manage to withstand the powers of the second one and subdue it.

In contemporary philosophy of action, Alfred Mele revitalized this discussion and established a modern account of self-control whereby self-control and weakness of will (akrasia) appear as the two sides of the same coin (Mele 1995). Those who are self-controlled overrule their unruly emotions, thoughts or urges, whereas those who are weak-willed give in to temptation and let their spontaneous responses reign. So far, the philosophical understanding of self-control seems well compatible with the models invoked in neuroscience and social psychology.

From a conceptual point of view, however, weakness of will is not the only opposite of self-control. In philosophical literature, weakness of will has typically been understood as intentional behaviour against one's better judgment (Davidson 1980). Formulated differently, people who act in a weak-willed fashion act freely (that is, they are not coerced) against what they consider best, or against their goal. This definition, however, has given rise to two further accounts of self-control failure which are typically termed *compulsion* and *recklessness,* respectively. Most basically, people can lack intentional control of their actions, in which case they act compulsively. They may, for example, want to give a presentation in front of their class, yet as soon as they see all the faces of their fellow pupils watching them, become so nervous that they can only stutter incoherently. Moreover, they may fail to take into consideration or pay due attention to their best judgment or the reasons that in their own eyes should guide their course of action. In case people fail to consider judgments that would otherwise have guided their action, people act recklessly. This would hold, for example, for a person who fully agrees that it would be best to prepare well for a presentation, but who still does not care and therefore does not prepare.

By drawing these distinctions, philosophical analysis intended to specify what exactly it means to act in a weak-willed fashion has actually given rise to a comprehensive three-tiered account of agential or self-control failure (e.g. Kennett 2001) distinguishing between compulsion, weakness of will and recklessness. A closer look at these three accounts of self-control failure can help us to understand a correspondingly comprehensive account of successful self-control.

The typical understanding of weakness of will presupposes that people do not only act intentionally while having the capacity to act as is required, but that they also hold a corresponding judgment and consider a certain goal worthwhile to achieve. Anyhow, people fail. In both compulsion and recklessness, one of these preconditions is not fulfilled. As said, compulsion denotes behaviour that is not intentional and hence in these kinds of failure the first precondition is not fulfilled. For the purpose of this chapter it is the failure of recklessness that is of particular importance. Recklessness seriously challenges the precondition of holding a best judgment by the very person – the self – who is supposed to act. Reckless people have formed a distinct judgment and hold certain goals; nonetheless they fail to take these as reasons for action. Formulated differently, they fail to care for their goals and guide their behaviour on the basis of what they themselves deem valuable and worthwhile. Conversely, successful self-control means that people care for their goals properly. That is, they consider these as providing them with reasons for action on the basis of which they alter and adapt their actual behaviour. Such an account of self-control, however, depends on a specific relationship agents have with the goals in question. Goals that are pre-defined or externally provided for are not even able to trigger behaviour that can count as an instance of *self*-control (in the sense of opposing recklessness).

The participants in the various experimental set-ups invoked in social psychology and neuroscience could not have failed recklessly. Either they succeeded, or they failed in a compulsive or weak-willed manner, conceptually speaking. However, once the possibility of reckless behaviour is taken seriously, the self's view of what counts as a reason for action becomes indispensable and it cannot be replaced by successful goal-achievement. In this sense, "the truly self-controlled agent, …, would intend and do what he judges he ought to do, because *he so judges*, and not because he has managed to skilfully circumvent or overcome a motivational state his judgment cannot control" (Campbell 2000: 116, my emphasis). By contrast, reckless people would somehow silence their own reasons or behave in a way that shows a disregard for the reasons actually available to them (Kennett 2001: 171ff). For example, if a devoted footballer or tennis player points out before skipping training that it is raining too much although it is only a light drizzle, she would be reckless. Such a person would, however, not be weak-willed. Being a devoted footballer, rather than one who only plays for the exercise or to please her parents, playing football provides her with a weighty reason to actually go and play. It is part of who she is and by not playing for any good alternative reason would embody a disregard of her own consideration, or in Kennett's words, she would "reach a judgment which is unfounded in the totality of reasons actually available" (p. 172) to her.

Understood this way, the prefix *self* does add meaning to the term control in the concept of self-control. Only in case a self's normative outlook of what counts as a reason for action for her is integrated in the respective behaviour, can the behaviour count as a *self*-controlled action. By contrast, only when a person has developed reasons for action – being a keen footballer – can she fail in a reckless way. In both cases, predetermined and externally set goals are excluded. This, however, seriously limits the practical implications of recent neuroscience and social psychology findings for raising self-controlled children. This is because the underlying experimental settings provide research participants with predefined goals rather than examine participants' view on what is worthwhile or a reason for action.

5.7.2 Self-Control as Agency

A closer look at philosophical discussions on the meaning of agency may shed further light on this issue and help elaborate on what it means when a self's perspective is involved in the behaviour rather than the goals that guide people's course of action. In philosophy of agency, an interesting distinction has been drawn between full actions and mere behavioural events (Wilson and Shpall 2012). Both concepts regard some behaviour as carried out by a specific person, yet the way in which the person is involved differs significantly. For behavioural events it can be said that the psychological and physiological processes are well in place so that successful goal-achievement can be safeguarded. However, for any behaviour to count as an action something more is required. The person has to be involved actively, so that she carries out the action as something which belongs to her personally, rather than that the behaviour is carried out due to certain psychological and physiological processes only.

The philosopher David Velleman (2000) pointed out this same point in a critique originally oriented at previous philosophical thought on the meaning of human behaviour. These discussions had centred on concepts of desire, intention, judgment, reason or bodily movement and investigated their respective interrelationship. Velleman, however, argued that in these standard theories of agency, specific causal relationships between these abstract concepts (e.g., reason causing intention and intention causing bodily movement) would have been considered sufficient to describe and explain human agency. The problem with this understanding of agency, however, is that "in this story, [...] nobody – that is no person – does anything. Psychological and physiological events take place inside a person, but the person serves merely as the arena for these events: he [or she] takes no active part" (Velleman 2000: 123). This, however, is what actually distinguishes full actions from mere events. Transferred into the terminology of self-control, I argue that only in case people are involved in their behaviour as agents, can their behaviour even be considered to be self-controlled in the sense of being controlled by the self. However, the question of whether the self as object is controlled as well is a different question and both questions can and have to be answered independently of each other.

Christine Korsgaard's distinction between what she called the combat-model versus the constitution-model of agency (Korsgaard 1999) can shed further light on this question. According to the combat-model, agents will continuously experience an inner conflict between some reason and some diverting passion. To put it into the terminology of empirical psychology and neuroscience, they would experience a conflict between some goal-state and some spontaneous response. Whenever reason prevailed, the result of the conflict would be a genuine (read: self-controlled) action, whereas in case passion prevailed the result would count as a mere behavioural event (read: lack of self-control). So the combat-model says. Hence, the suggestion is that whenever reason or some goal prevailed, corresponding behaviours were instances of agency or self-control.

The problem with this understanding of agency, or self-control, Korsgaard says, is that it cannot explain why reasons, or goals, are superior to desires, or spontaneous responses. As a consequence, the combat-model cannot provide room for people to evaluate and determine the relative weight or relevance of desires and eventually to adjust their reason or goals in case they consider these to be inadequate or even undesirable. A constitution-model of agency, however, could make good for this, so Korsgaard argues. According to this model, agency is in place only and because the person's constitution, that is who she is, is involved. In that case, people live up to reason, because it is *their* reason or because *they* consider the goal worthwhile to achieve. Analogously, one could argue that whenever a person deliberates about the desirable course of action, or about whether or not to adapt her current behaviour, the person who adapts her behaviour is inevitably involved in such decisions. A desire may be so strong that it outweighs the force of a reason, resulting in behaviour that may be best described as an instance of weakness of will.

According to Korsgaard, a different understanding of the situation is also available. "The strength of a desire may be counted by you as a reason for acting on it; but this is different from its simply winning" (Korsgaard 1989: 370). For example, a keen footballer may decide that her dislike of playing football during a severe storm may be a good reason not to play and stay at home. In that case, the person acts on the basis of the desire [dislike to go outside], yet in case she considers it appropriate [the storm is too heavy to go outside] and a reason for action for herself, acting accordingly counts as an instance of successful agency rather than an event or a failed action. So the agential value of any behaviour carried out by a specific person is mainly determined by the question of whether and how her constitution is involved. Goal-achievement [be on the football pitch and play] and response-suppression [go outside despite dislike] as such are not decisive criteria of whether or not a person exercises self-control. Instead, self-control is on hand if and because a person considers a certain goal-state to be appropriate and worthwhile and as something that she ought to achieve.

5.8 Raising Self-Controlled Children Revisited

The above analyses present rich information from recent research findings in neuroscience, social psychology and philosophical work on the meaning of the concept of self-control. In this final paragraph I reflect on this and ponder answers to the initial question presented in this chapter. What would a parental responsibility to raise self-controlled children entail and how does or should recent neuroscience and social psychology inform this responsibility? Halfway through the chapter the answer seemed straightforward and clear. Yet, after having reflected on the concept in greater detail, issues appear more complex and easy answers are no longer convincing.

The chapter started off by showing that self-control denotes people's ability to alter their actual behaviour in order to achieve certain ends or, formulated differently, to prevent and override unwanted thoughts, behaviours and emotions. In light of such definitions, experimental research in both neuroscience and social psychology mainly focused on means and mechanisms related to the alteration of behaviour and the prevention and suppression of immediate responses. The goals or the target-states, however, are hardly picked out as separate issues; instead they are typically predefined and given to research participants. Thereby, neuroscience and social psychology research on self-control is silencing essential aspects of its own topic. The research communities here presuppose a self rather than investigate it as the subject of behavioural control. However, as self-determination theory indicated and recent philosophical work argued, a profound understanding of the self is crucial in order to grasp what self-control means and what is required for self-controlled behaviour. Explicit attention to the self as the subject of self-controlled behaviour is required to determine the appropriateness and desirability of specific goals, as well as, the inadequacy of potentially opposing urges, thoughts or behaviours.

From an educational perspective, this double background according to which self-control is both control of the self and control by the self provides a variety of possibilities. Two different sets of educational goals become available that relate to the task of raising self-controlled children. One of these sets regards the enablement of children to control themselves so they can achieve some distant or desirable goal. This set of educational goals includes goals like obedience, accepting and following externally-set rules, or the suppression of impulses and immediate desires. The other set, however, regards the processes that support children in developing a self in the sense of becoming able to set goals for themselves and to determine what they consider desirable or worthwhile (Horstkötter and Snoek 2013). This second set of educational goals requires reflection on the normative value of possible goals, evaluating the desirability of one's actual behaviour or feelings, or the development of a self who sets goals and considers these then as self-evident reasons for her action.

I argued that in order to raise self-controlled children, parents can teach them established self-control skills and strategies, train their self-control strength, or enrol them in self-control testing and intervention programmes. By doing this, parents could realise the first set of educational goals aimed at a capacity to control the self, according to some pre-set goals. However, these activities may not help parents in also achieving the second set of educational goals. To that end, parents are required to aim at enhancing the capacity for autonomy in children (e.g. Betzler 2015) or to contribute to the development of the 'subjective self' of their children, as it has been argued recently (Wringe 2015).

Supporting their children in becoming autonomous agents should then be considered as a distinct way to safeguard their long-term wellbeing. Therefore, raising self-controlled children is a comprehensive undertaking inevitably connected to two different sets of educational goals. It denotes activities that enable children to suppress their immediate desires and to delay gratifications and it covers endeavours that contribute to the development of a self and a normative outlook that teaches children to determine when to suppress which responses and why to aim at which goals. Parents' first responsibility would then be to acknowledge the comprehensiveness of raising self-controlled children. In so far as self-control contributes to children's wellbeing, it is important to further investigate how parents can live up to the two sets of educational goals highlighted. Against this background, the relevance of recent neuroscience and social psychology research may be more limited than it appeared at the beginning of and halfway through this chapter. While research findings can be useful to inform parents about what to do to instil in their children the capacity to suppress spontaneous responses, these findings alone cannot determine the value for their children of certain distant goals or spontaneous desires. In this sense, while being insightful and informative, current neuroscience and social psychology alone are not sufficient to raise self-controlled children.

References

Aha! Parenting. 2013. 8 steps to help your child develop self-control. Retrieved from www.ahaparenting.com/_blog/Parenting_Blog/post/Help_Your_Child_Develop_Self_Control/.

Aristotle. 1980. *The Nicomachean ethics*. Trans. D. Ross. Oxford: Oxford University Press.

Aron, A.R., T.W. Robbins, and R.A. Poldrack. 2004. Inhibition and the right inferior frontal cortex. *Trends in Cognitive Science* 8(4): 170–177.

Baumeister, R.F. 2012. Self-control the moral muscle. *The Psychologist* 25(2): 112–115.

Baumeister, R.F., and T.F. Heatherton. 1996. Self-regulation failure: An overview. *Psychological Inquiry* 7(1): 1–15.

Baumeister, R.F., and J. Tierney. 2011. *Willpower, why self-control is the secret to success*. London: Allan Lane.

Baumeister, R.F., T.F. Heatherton, and D.M. Tice. 1994. *Losing control: How and why people fail at self-regulation*. San Diego: Academic.

Berkman, E.T., A.M. Graham, and P.a. Fisher. 2012. Training self-control: A domain-general translational neuroscience approach. *Child Development Perspectives* 6(4): 374–384.

Betzler, M. 2015. Enhancing the capacity for autonomy: What parents owe their children to make their lives go well. In *The nature of children's wellbeing, theory and practice*, ed. A. Bagattini and C.M. Macleod, 65–84. Berlin/New York: Springer.

Browers, E. 2012. *The everything's parent's guide to the strong-willed child, A positive approach to increase self-control, improve communication and reduce conflict*. Avan: Adams Media.

Campbell, P.G. 2000. Diagnosing agency. *Philosophy, Psychiatry and Psychology* 7(2): 107–119.

Carver, C.S., and M.F. Scheier. 1998. *On the self-regulation of behavior*. Cambridge, MA: Cambridge University Press.

Casey, B.J., L.H. Somerville, I.H. Gotlib, O. Ayduk, N.T. Franklin, M.K. Askren, et al. 2011. Behavioral and neural correlates of delay of gratification 40 years later. *Proceedings of the National Academy of Sciences* 108(36): 14998–15003.

Cervone, D., W.G. Shadel, R.E. Smith, and M. Fiori. 2006. Self-regulation: Reminders and suggestions from personality science. *Applied Psychology: An International Review* 55(3): 333–385.

Davidson, D. 1980. How is weakness of the will possible? In *Essays on actions and events*, ed. D. Davidson, 21–42. Oxford: Clarendon.

Deci, E.L., and R.M. Ryan. 2000. The "what" and "why" of goal pursuit: Human needs and the self-determination of behavior. *Psychological Inquiry* 11(4): 227–268.

Frankfurt, H.G. 1988. Freedom of the will and the concept of a person. In *The importance of what we care about*, ed. H.G. Frankfurt, 11–25. Cambridge, MA: Cambridge University Press.

Gottfredson, M.R., and T. Hirschi. 1990. *The general theory of crime*. Stanford: Stanford University Press.

Hagger, M.S., C. Wood, C. Stiff, and N.L.D. Chatzisarantis. 2010. Ego depletion and the strength model of self-control: A meta analysis. *Psychological Bulletin* 136(4): 495–525.

Hare, T.A., C.F. Camerer, and A. Rangel. 2009. Self-control in decision-making involves modulation of the vmPFC valuation system. *Science* 324: 646–648.

Henden, E. 2008. What is self-control? *Philosophical Psychology* 21(1): 69–90.

Horstkötter, D. 2015. Self-control and normativity. Theories in Social Psychology Revisited *Theory and Psychology* 25(1): 25–44.

Horstkötter, D., and A. Snoek. 2013. The importance of the self for autonomous behavior. *American Journal of Bioethics Neuroscience* 4(4): 62–63.

Kennett, J. 2001. *Agency and responsibility: A common-sense moral psychology*. Oxford: Clarendon.

Korsgaard, C.M. 1989. Personal identity and the unity of agency: A Kantian response to Parfit. *Philosophy and Public Affairs* 18(2): 101–132.

Korsgaard, C.M. 1999. Self-constitution in the ethics of Plato and Kant. *The Journal of Ethics* 3: 1–29.

Loeber, R., B. Menting, D.R. Lynam, T.E. Moffitt, M. Stouthamer-Loeber, R. Stallings, and D. Pardini. 2012. Findings from the Pittsburgh Youth Study: Cognitive impulsivity and intelligence as predictors of the age-crime curve. *Journal of the American Academy of Child and Adolescent Psychiatry* 51(11): 1136–1149.

McGonigal, K. 2012. *The willpower instinct. How self-control works, why it matters and what you can do to get more of it*. New York: Penguin.

Mele, A.R. 1995. *Autonomous agents: From self-control to autonomy*. Oxford: Oxford University Press.

Mischel, W. 2014. *The marshmallow test: Mastering self-control*. New York: Little, Brown and Company.

Mischel, W., E.B. Ebbesen, and A. Raskoff Zeiss. 1972. Cognitive and attentional mechanisms in delay of gratification. *Journal of Personality and Social Psychology* 21(2): 204–218.

Mischel, W., Y. Shoda, and M.L. Rodriguez. 1989. Delay of gratification in children. *Science* 244: 933–938.

Mischel, W., N. Cantor, and S. Feldman. 1996. Principles of self-regulation: The nature of willpower and self-control. In *Social psychology: Handbook of basic principles*, ed. E.T. Higgins and A.W. Kruglanski, 329–360. New York: Guilford Press.
Moffitt, T.E., L. Arseneault, D. Belsky, N. Dickson, R.J. Hancox, H. Harrington, and A. Caspi. 2011. A gradient of childhood self-control predicts health, wealth, and public safety. *Proceedings of the National Academy of Sciences* 108(7): 2693–2698.
Muraven, M., and R.F. Baumeister. 2000. Self-regulation and depletion of limited resources: Does self-control resemble a muscle? *Psychological Bulletin* 126(2): 247–259.
Muraven, M., D.M. Tice, and R.F. Baumeister. 1998. Self-control as limited resource: Regulatory depletion patterns. *Journal of Personality and Social Psychology* 74(3): 774–789.
Plato. 1952. *Phaedrus*. Trans: R. Hackforth. Cambridge, MA: Cambridge University Press.
Ryan, R.M., and E.L. Deci. 1999. Approaching and avoiding self-determination: Comparing cybernetic and organismic paradigms of motivation. In *Perspectives on behavioral self-regulation*, vol. 12, ed. R.S. Wyer Jr., 193–215. Mahwah: Lawrence Erlbaum Associates.
Ryan, R.M., and E.L. Deci. 2002. An overview of self-determination theory: An organismic-dialectical perspective. In *Handbook of self-determination research*, ed. E.L. Deci and R.M. Ryan, 3–33. Rochester: The University of Rochester Press.
Slutske, W.S., T.E. Moffitt, R. Poulton, and A. Caspi. 2012. Undercontrolled temperament at age 3 predicts disordered gambling at age 32: A longitudinal study of a complete birth cohort. *Psychological Science* 23(5): 510–516.
Strayhorn, J.M.J. 2002. Self-control: Toward systematic training programs. *Journal of the American Academy of Child and Adolescent Psychiatry* 41(1): 17–27.
Velleman, D.J. 2000. What happens when someone acts? In *The possibility of practical reason*, ed. D.J. Velleman, 123–143. Oxford: Clarendon.
Wilson, George, and Shpall, Samuel. 2012. Action, In *The Stanford encyclopedia of philosophy* (Summer 2012 Edition), ed. Edward N. Zalta, URL = http://plato.stanford.edu/archives/sum2012/entries/action/.
Wringe, C. 2015. Beyond useful knowledge: Developing the subjective self. *Journal of Philosophy of Education* 49(1): 32–44.

Chapter 6
Direct Electrical Stimulation of the Developing Brain: Who Decides?

Farah Focquaert

Abstract In this chapter I focus on the ethics of direct stimulation of the developing brain via electrodes and the parental responsibilities that are actualised by the possibilities brought about by invasive therapeutic research for children. I specifically address the ethics of paediatric decision-making and participation in the context of investigational treatments. I present an updated account of my ethical view on paediatric deep brain stimulation and identify parents' decision-making responsibilities within the context of invasive investigational treatments such as deep brain stimulation. It is generally accepted that parents have a right to decide on behalf of their children – this right is dependent on the assumption their choices will be made in the best interests of the children. However, we have reason to doubt the claim that parents always or generally act in the best interests of their children. I argue that a shared decision-making approach more effectively protects children's rights, secures the best possible care and support during and after the intervention, preserves family intimacy and stimulates the developing autonomy of the child patient. I conclude that parental responsibility entails that parents should adopt a shared decision-making approach regarding interventions like direct stimulation of the developing brain.

Keywords Children • Deep brain stimulation • Direct electrical stimulation • Neuromodulation • Paediatric decision-making • Assent • Dissent • Consent • Parental authority • Shared decision-making • Autonomy • Developing autonomy • Best interest

---

F. Focquaert (✉)
Bioethics Institute Ghent, Department of Philosophy and Moral Sciences, Ghent University, Blandijnberg 2, Ghent 9000, Belgium
e-mail: farah.focquaert@ugent.be

K. Hens et al. (eds.), *Parental Responsibility in the Context of Neuroscience and Genetics*, International Library of Ethics, Law, and the New Medicine 69,
DOI 10.1007/978-3-319-42834-5_6

6.1 Introduction

Deep brain stimulation (DBS) involves direct stimulation via electrodes in areas located deep inside the brain, and may be therapeutically indicated as a last intervention in children and adolescents with severe and/or treatment-refractory movement disorders such as dystonia. Dystonia is a hyperkinetic neurological movement disorder characterized by involuntary twisting and turning movements and abnormal posturing. In severe cases it can lead to fixed orthopaedic deformities. Although quite rare, dystonia is the most frequently encountered movement disorder in children. It comes in many different forms[1] and can be triggered by various factors encompassing genetic, neurodegenerative, metabolic, inflammatory and other causes. In primary dystonia[2] all symptoms are caused by the dystonia itself, while in secondary dystonia the symptoms result from a clearly identified and frequently acquired neurological disorder (e.g., cerebral palsy due to focal brain lesions). Most of the patients with primary dystonias do not suffer from cognitive deficits, while patients with secondary dystonias may experience mild to severe cognitive deficits. Patients with generalized dystonia, affecting the entire body, are often severely impaired and wheelchair-bound, but patients with isolated dystonia, restricted to a specific body part or region such as the mouth, neck, arms or legs, may also face considerable hindrance and suffering.

More than 100,000 individuals have received DBS treatment worldwide. DBS is standard last-resort treatment for medication-refractory Parkinson's disease and essential tremor, for which it has been used most frequently. Moreover, the United States Food and Drug Administration (FDA) approved a humanitarian exemption for obsessive compulsive disorder (OCD) in 2008. A humanitarian device exemption removes the requirement for a clinical trial of the appropriate size and statistical power for medical devices (Fins et al. 2011). During the last decades, DBS via electrode implants has been used more frequently in a variety of investigational applications, including treatment refractory disorders such as Tourette syndrome (TS), depression, anxiety, cluster headache, obesity, chronic aggression, minimally conscious state, tinnitus, addiction and even Alzheimer's disease (Unterrainer and Oduncu 2015). In paediatric patients, DBS has been used most commonly for dystonia. Although paediatric DBS is investigational at present, DBS for dystonia has been FDA approved in 2003 under a humanitarian device exemption in adults and children of 7 years old and older (DiFrancesco et al. 2012). Children as young as four have been implanted for dystonia (Kaminska et al. 2012; Starr et al. 2015).[3]

[1] Focal: single body region, segmental: more than one body region, generalized: two or more contiguous body parts and torso, hemidystonia: one side of the body, multifocal: non-contiguous body regions.

[2] Primary dystonia is either genetically caused or has an unknown cause.

[3] In line with the humanitarian device exemption status, pediatric DBS must be performed in a medical center whose institutional review board (IRB) has approved use of the device. Pediatric DBS for dystonia also received CE-mark approval for its use in Europe in 2013, which entails that the procedure meets the requirements of the EU Directive 90/385/EEC regarding active implant-

There are case reports of DBS for psychiatric disorders in adolescents (e.g. TS in 16- and 17 -year olds), but not in very young children to the best of my knowledge. DiFranscesco and colleagues (2012) discuss childhood dystonia, TS, OCD, obesity, epilepsy and juvenile Parkinson's disease (PD) as potential paediatric indications for DBS. Whereas dystonia is the most promising paediatric application of DBS at present due to the low remission rate and significant disability if left untreated, DiFrancesco and colleagues caution the application of DBS for paediatric conditions with a high remission rate in adulthood such as TS and OCD.

A case series (Air et al. 2011) of DBS in paediatric patients with primary generalized dystonia showed excellent results, especially in paediatric patients without fixed orthopaedic deformities. Results in secondary dystonia appear to be less successful and much more variable. Overall, data on DBS in paediatric movement disorders is very limited (i.e., single case reports or small case series). Moreover, predictive factors for DBS outcome are not very well known for primary and secondary dystonia and a much higher incidence of hardware (batteries, leads, or extension cables) infections has been reported in paediatric patients (13 %) compared to adults (4–8 %).[4] All the infections occurred in patients younger than 10 years of age, with severe disability and a diagnosis other than primary dystonia (Air et al. 2011). Also, brain growth, especially before 7 or 8 years of age and during the growth spurt in puberty, may cause displacement in electrodes (Woopen et al. 2013). Nonetheless, several isolated cases and small case series seem to indicate that earlier intervention (e.g., <16) is more beneficial compared to later intervention, and that a shorter duration of symptoms (in terms of months and years) and less severe symptoms (e.g., no fixed orthopaedic deformities) seems to be associated with better treatment outcomes (Woopen et al. 2013). Also, Kaminska et al. (2012) report a low complication rate in a series of paediatric patients implanted with rechargeable deep brain stimulators. Importantly, effects of early DBS can be significant in terms of motor skills, cognitive function, completion of education, and social development. As Woopen and colleagues (2013) argue:

> Children with severe movement disorders like CP [cerebral palsy] are considerably impaired in their social abilities to maturate in a normal environment: Schooling, meeting friends, making sports, etc. is often extremely difficult for them. The impact of a considerable improvement and the resulting increase in mobility and autonomy can, in our view, not be overestimated (83).

---

able medical devices and must follow the Declaration of Helsinki (WMA 2013), including the need for an investigational protocol and IRB oversight and approval.

[4]According to an expert team performing DBS for movement disorders in adults, the most serious risk of the surgical procedure is bleeding producing a stroke, which occurs in 1–3 % of cases and may be life-threatening. Infection of the deep brain stimulation hardware occurs in 4–8 % of cases and is usually not life-threatening (Air et al. 2011; Starr et al. 2015). If serious and/or unwanted behavioral (cognitive, motor, emotional, etc.) side effects occur due to stimulation, the device can be switched off permanently without the need to remove the intracranial hardware. Removing the electrode may pose risks in terms of tissue damage and bleeding.

This chapter focuses on the ethics of direct stimulation of the developing brain[5] via electrodes such as deep brain stimulation. I specifically address the ethics of paediatric decision-making and participation in the context of investigational treatments. Previously, I have argued that children and adolescents should not *by default* be excluded from receiving interventions such as DBS which involve brain surgery and direct stimulation of the brain via electrodes (Focquaert 2011). Harm may result from the unwillingness to relieve severe, chronic suffering in those cases where an investigational treatment is the only option left. I further argued that shared-decision making in the case of paediatric DBS, between the parents and the child patient and in deliberation with the medical team, provides the strongest safeguards for the child patient in a variety of ways (e.g., protecting the developing autonomy of the child, protecting family intimacy, better treatment outcomes, etc.) (Focquaert 2013). I also argued that, due to the invasiveness and investigational nature of the procedure, paediatric DBS should only be considered if successful treatment outcomes have been obtained in adults (Focquaert 2011).

In this chapter, I will zoom in on both ethical issues. I review why shared decision-making can be considered the most ethically defensible path in paediatric DBS, and briefly discuss if we should indeed wait for successful treatment outcomes in adults before moving on to paediatric patients. Based on the comments of Woopen et al. (2013), I present an updated account of my ethical view concerning direct electrical stimulation of the developing brain and identify parents' decision-making responsibilities within the context of invasive investigational treatments such as DBS. In general, the medical team's primary responsibility is to safeguard the child's interest when parents bring her for treatment. Parents are generally assumed to aim to promote and protect their children's wellbeing (see Björnsson & Brülde and Syurina & Feron in this volume). Parents have the right to consent to treatment on behalf of their (incompetent or not yet competent) children provided that the treatment is in the best interest of the children. Of course, parents' right to decide for their children has limits and parents are typically not morally and legally allowed to refuse medically appropriate treatment which is in the child's best interests (e.g., parents who are Jehovah's witnesses are not allowed to prevent their children from receiving life-saving blood transfusions), nor request inappropriate treatment for their children. According to The British Medical Association (2008), the "moral authority behind parental responsibility depends in large part on the entirely reasonable supposition that parents will act in the best interest of their children". I argue that we have reason to doubt the claim that parents always or generally act in the best interest of their children. Based on the analysis in this chapter, I therefore conclude that parental decision-making responsibility entails that parents

[5] In academic neurological journals and the medical world often the term 'paediatric brain' is used. I, however, prefer to use 'developing brain' here as the term paediatric refers to a branch of medicine, that is medicine for children where 'paediatric' is made of words referring to both the child ('pedo') and the healer ('iatros'). Hence, a brain itself cannot be paediatric, instead the brains of children are developing.

should adopt a shared decision-making approach when interventions like DBS are considered.

6.2 Paediatric Decision-Making in Therapeutic Research Settings

Whether and to what extent children should be included in the decision-making process is a very serious ethical issue that needs to be fleshed out if children and adolescents are to be exposed to the risks involved in therapeutic DBS research. If children are not given an active role in the decision-making process, they may become victims of deception, abuse or coercion by third-party decision-makers (e.g., parents) when the decision to intervene is mediated by caregiver interests rather than the child's best interests (Focquaert 2013). Moreover, according to Woopen et al. (2013), if direct stimulation of the brain is considered to be the appropriate therapeutic approach, and children have no decision-making powers, then they cannot claim an intervention or investigational treatment even if it would be medically indicated within the context of (last-resort) investigational treatments. Hence, children may be the victims of interventions that are initiated 'too early' or 'too late' from a therapeutic perspective (Woopen et al. 2013).

Whereas the parents' role in medical decision-making is typically very straightforward in legal and ethical guidelines on paediatric research (e.g. within research contexts parents give consent on behalf of their children), the child's role is not always clearly defined. Children's inclusion in the informed consent process, or the need to receive their assent in addition to the parents' consent, is often framed as dependent on their capacities (e.g., ability to retain information and form their own opinion). With respect to the ethical issues, Woopen and colleagues (2013) point to the common intuition that there are relevant differences between children's and adults' capacities to understand the disease, the proposed intervention and its consequences, and the impact it may have on the child's future life.

At present, paediatric DBS for dystonia is considered an investigational treatment and falls into the category of therapeutic research. Therapeutic research involves experimental case series or clinical trials in which the intervention under investigation is believed to likely benefit the participants receiving the experimental drug or procedure. "Investigational treatment" means research involving drugs or procedures that has undergone basic laboratory testing and received FDA approval to be tested in humans (US) or marketing authorization for its use in other diseases or conditions (Europe). It is experimental in nature, but involves a drug or procedure for which effective outcomes and an acceptable risk/benefit ratio have been established in specific other diseases or conditions.[6] Hence, international and national

[6] It is therefore different from 'innovative medicine' where new procedures can be tested without IRB oversight and approval. Due to the humanitarian device exemption status in the US and the CE-mark approval requirement in Europe, pediatric DBS for dystonia must be performed in a

laws regarding research on humans should be followed and the international ethical guidelines concerning research on humans should be respected. For example, according to the Belgian law regarding research on humans (W 2004-05-07/32), the refusal of minors who are deemed competent to give their assent/dissent should be respected by the investigator. Hence, although not explicitly mentioned, paediatric patients have a veto right and their refusal to participate in any kind of investigational treatment should be respected.

According to the Declaration of Helsinki (WMA 2013), if a research participant is deemed incapable of giving informed consent (e.g., because she does not yet have this capacity), the physician must seek the assent of the participant in addition to the consent of the legally authorized representative (e.g., parents). Importantly, the *dissent* of the potential participant must be respected. This means that the child has a veto right. If the parents consent and the child dissents, then the child should not be enrolled in a research study or be subjected to an investigational treatment such as paediatric DBS. According to the Regulation EU no. 236/2014 201 of the European Parliament and the Council of Europe concerning clinical trials, "the minor shall take part in the informed consent procedure in a way adapted to his or her age and mental maturity" and the "explicit wish of a minor who is capable of forming an opinion" to refuse participation in the research trial or investigational treatment, or to withdraw from it, should be respected by the investigator. Hence, the child's dissent should be respected and a dissenting child should not be enrolled in a research study or be subjected to an investigational treatment.

Can children be competent enough to decide on an intervention such as DBS? On average, children's autonomy increases with age, life experiences, and their active involvement in decision-making as a learning process. Children's unique personality, life experiences (e.g., having experienced a serious disease for many years) and education may considerably impact their decision-making skills regardless of their age. Many children with movement disorders are only mildly cognitively impaired or may even have normal cognitive abilities. These paediatric patients are perfectly able to be adequately informed about the DBS procedure and give their assent or dissent. However, some children may have severely impaired communicative abilities or suffer from severe mental impairment and may not be able or competent to assent or dissent (Woopen et al. 2013). For those children, if DBS is therapeutically indicated, the parents should give their informed consent before the investigational treatment can be initiated, while the child should participate in the decision-making process as much as possible.

---

medical center whose IRB has approved use of the device. Only highly skilled, interdisciplinary DBS research teams possess the much needed expertise to perform DBS in new diseases or conditions with sufficient safety and caution. The invasiveness of the procedure demands IRB oversight and approval for all new applications. New DBS applications (e.g., for pain or psychiatric diseases) are therefore typically performed within a therapeutic research context involving IRB oversight and approval.

6.3 Shared Decision-Making

Woopen and colleagues (2013) state that in the case of choosing DBS over medication therapy it is "the patient himself who has to weigh the respective advantages and disadvantages of each procedure according to his needs, values and preferred way of living" (Woopen et al. 2013: 80). Should this also be the case in paediatric patients? Should we respect children's and adolescents' developing autonomy as much as possible and grant them real decision-making powers, involving a veto right with respect to their parent's decisions, or should the invasiveness of the procedure direct us away from shared decision-making, where parents and children take decisions together, and towards parental authority?

In Focquaert (2013), I specifically addressed the pros and cons of the two most commonly discussed approaches to decision-making in the case of paediatric patients: parental authority in decision-making on the one hand, and shared-decision making on the other hand. DBS is an invasive procedure because it involves brain surgery, the placement of electrodes inside one's brain and of a battery under the skin near the collarbone or in the thigh area. A variety of mild to severe side effects related to the surgery, the technical devices, as well as the stimulation, may occur. Hence, one may be inclined to argue that due to the invasiveness of the procedure and the possibility of death due to surgical complications, parents should have sole decision-making power except in those cases where the child can be considered a competent minor.

One might argue that children are cognitively and emotionally incapable of deciding for themselves and of exercising their rights. 'Child protectionists' emphasize that too much weight has been placed on children's autonomy and argue that respecting children's autonomy in certain cases may be detrimental (e.g., because children may not possess the same ability as adults to take their future life plans into account) rather than empowering (Alderson 1992). Ross (2004) has argued that respect for children not only implies respect for their current autonomy, but also, and perhaps even more so, respect for their future autonomy and the person they are becoming. She states that parents are able to take their child's current and future autonomy into account, by focusing on likely short-term and long-term consequences of certain decisions, whereas children typically do not possess such capacities or at least not to the same degree. According to Ross (2004), parents respect their children by deciding *for* them, rather than by giving them an actual say in the decision-making process. Parental decision-making authority is justified because (a) parents care deeply about the welfare of their children and know best who they are and what their needs are, (b) parents bear the burdens of their children's healthcare treatment (e.g., safeguard postoperative care), (c) parental authority promotes family intimacy, and because (d) parents have, within limits, the right to raise their children according to their values and standards.

Of course, parents are morally required to make decisions in their child's best interests. However, are parents always motivated by their children's best interests and do parents have privileged access to and knowledge about their children's current and future bests interests, their strengths and weaknesses? Although child protectionists claim they are and they do, this is not always the case. Parents may regularly breach their obligations towards their children. Child maltreatment (e.g., physical abuse, sexual abuse, psychological abuse and neglect) data clearly show that parents do not always act in their child's best interests. Child maltreatment has been described as a common phenomenon in high-income countries and, except for sexual abuse, 80 % or more of maltreatment is perpetrated by parents (Gilbert et al. 2009). Moreover, Hart and Chesson (1998) warn us that one should not assume that parents' reflection of their child's illness is an accurate reflection of the child's actual state of mind. Research has shown that children tend to protect their parents; that adults project their own experience of the child's illness onto their children; and that divergent opinions exist between parents and medical staff concerning the child's state of mind. Moreover, a number of studies have found that self-reported quality of life often differs from proxy reports, both when considering adult patients and child patients (see Focquaert 2013). In sum, parents do not always know what is in their child's best interests and do not always act according to their child's best interests. 'Child liberationists' therefore argue that parents have the duty to gradually diminish their proxy decision-making and allow their children to make their own choices in order to stimulate their developing autonomy. If children are not allowed to participate in decisions regarding their own life, how will they acquire the necessary knowledge and capacities that will enable them to make informed decisions that promote their life goals once they are adults? By providing opportunities to choose for themselves, children can develop and learn the necessary skills to become fully competent adults (Hagger 2009).

If one claims that children cannot make their own choices because they lack the necessary experience, then denying them of all opportunities to choose basically prevents them from ever acquiring it. Although the invasiveness of direct electrical stimulation of the brain may count against child participation in the decision-making process, the same rationale may promote active participation on behalf of the child. The more impact a procedure can have on the current and future autonomy of the child, the more the vulnerable child patient may be impacted by the wrong decision. This is especially so in case of DBS for conditions that have an uncertain disease prognosis if left untreated and symptoms may improve or spontaneously remit with time (e.g., TS, OCD). To maximize children's autonomy, Hagger (2009) suggests to avoid arbitrary cut-off points (such as age) in medical decision-making where possible. If there is disagreement concerning a particular treatment option, or if the medical team suspects that the child is assenting under duress, then a psychological assessment of the child's decision-making capacities and of the voluntariness of her decision should be made on a case-by-case basis. If there is agreement between the parents and the child and if the medical team believes that the intervention is in the child's best interests and that the child's assent is meaningful and unforced, then there's no need to further assess the child's decision-making skills. From a weak or conservative 'child liberationist' perspective, rather than enforcing child autonomy, parents and parental guardians should have adequate respect for children's rights where appropriate, meaning that the context and the child's life experiences and cognitive abilities are crucial in determining the active involvement of the child, rather than necessarily granting children decision-making rights in all areas of life (Alderson 1992; Focquaert 2013).

According to Woopen and colleagues (2013), "intervening into the brain is like intervening into the midst of the person and his identity" (80). Clearly, this is no different for adults than it is for competent minors. If we understand identity as narrative identity (Schechtman 2010), then both minors and adults may have the ability to cohesively incorporate life events, even dramatic events, into their identity, or to reject certain changes if they consider these to be too disruptive of their character, desires, aspirations and goals. For example, whereas DBS for PD often results in marked improvement in motor symptoms, several patients experience stimulation-induced cognitive, emotional and motivational side effects. Severe psychological effects such as clinical disinhibition, addiction and aggression may arise. Some individuals may experience such acute and dramatic changes in their psychological make-up that it frightens them, and leaves them feeling alienated and without a sense of control and agency. For example, Leentjens and colleagues (2004) describe a case study of an individual with Parkinson's disease who became euphoric and disinhibited after DBS surgery. The 62 -year old man entered a relationship with a married woman and started spending huge amounts of his savings on clothing and partying. His disinhibition and megalomania gradually worsened into a manic disorder. Schechtman (2010) gives the example of a depressed person who was overwhelmed by the abruptness of his psychological changes: "The psychological changes brought about are so profound and occur so quickly that they can seem to break off one narrative – the story of a depressed person – and start a new one – that of a happy person" (137). Moreover, several individuals receiving DBS for PD experience mild forms of irritability, impatience and distractions which often disrupt existing relationships, aggravate pre-existing marital problems and frequently lead to familial conflicts and adjustment problems at work (Schüpbach et al. 2006). Contrary to DBS for PD, DBS for dystonia has not been associated with such common psychological side-effects (DiFrancesco et al. 2012). Future paediatric indications may nevertheless result in psychiatric side-effects, and this possibility should be clearly communicated and monitored in both adult and child patients.

The starting point in case of research with paediatric patients capable of giving their assent/dissent should always be a scientifically sound hypothesis of benefit versus risk of damages and burdens (Woopen et al. 2013). General knowledge of the side effects and possible adverse effects of DBS have to be weighed against the severity of suffering and the efficacy (or absence) of other treatment options. One of the most straightforward cases in terms of risk benefit analysis is early intervention for primary dystonia. Early intervention in carefully selected paediatric patients with primary dystonia (a) prevents irreversible damage such as orthopaedic deformations, (b) ensures the best outcome since the prognosis gets worse with severity and duration of the symptoms, and (c) prevents dramatic psycho-social effects due to social isolation, while maximizing motor learning and development (Focquaert 2011). As Woopen and colleagues (2013) argue, for a variety of other paediatric conditions, "a difficult balance of avoiding harm by preventing complications and alleviating suffering at an early age on one side, and taking the risks and burdens of DBS on the other side is necessary" (84) on a case-by-case basis. Moreover, as I have argued elsewhere (Focquaert 2011), children's benefits precede those of others and performing paediatric DBS solely to relieve caregiver burden is not ethically defensible. If DBS treatment is performed for childhood TS or OCD that might have spontaneously remitted or become subclinical with time, then the dissenting child patient is harmed because an unnecessary invasive procedure was forced upon her, and the only benefit that occurred is a third-party benefit (i.e., caregiver relief).

If children dissent in cases where DBS is strongly therapeutically indicated, it is important to try to understand why the child dissents (Focquaert 2011; Woopen et al. 2013). In those cases a professional psychologist can be consulted to talk with the child, discuss the situation the child is facing and potentially alleviate unwarranted fears (e.g., irrational fear of needles or hospital environments). Whether or not children's dissent may be overridden in cases where irreversible damage is likely if left untreated and therapeutic benefit can be expected based on scientific data is a difficult ethical question. For example, in extreme cases of severe primary dystonia where severe harm can be expected both in terms of orthopaedic deformities and social isolation, it may be ethically warranted to override a child's dissent. However, in cases where the disease prognosis is unclear and/or the condition may spontaneously remit by later adulthood, it is not ethically acceptable to override a child patient's dissent. Forcing a child patient to undergo DBS for a condition that would have spontaneously remitted if left untreated harms the child patient by unnecessarily subjecting her to the risks and burdens of the intervention (Focquaert 2011).

In cases where paediatric DBS is therapeutically indicated, Woopen et al. (2013) and Unterrainer and Oduncu (2015) follow my previous argumentation (Focquaert 2011; 2013) in recommending a shared decision-making process (i.e., dual consent procedure) for (early) DBS in paediatric patients. The parents should give their consent and the child her assent. They agree that a shared decision-making approach maximally protects children's rights and secures the best possible care and support during the treatment process, preserves family intimacy and stimulates the developing autonomy of the child patient.

6.4 Adults First

Woopen and colleagues (2013) discuss whether scientific proof of success in adults is a prerequisite for applying DBS in children and conclude that this claim cannot be substantiated. I agree that this claim is indeed scientifically and ethically questionable. Although I previously argued that proof of success in adults is required before enrolling child patients in investigational DBS settings due to the invasiveness of the procedure and the vulnerable nature of child patients, I also argued that the vulnerable nature of child patients should not by default exclude them from receiving the only intervention that could alleviate their suffering (Focquaert 2011).

The vulnerable nature of child patients should not be used as a rationale to prevent their participation in carefully selected investigational treatments if excluding them from such interventions may make them even more vulnerable. Excluding paediatric patients from receiving the only intervention that could potentially alleviate their extreme suffering may harm rather than protect vulnerable child patients. The same line of reasoning may justify the participation of certain child patients in investigational DBS even if success rates are not established in adults. Moreover, it is important to remember that direct stimulation of the brain in adults may have different effects compared to direct stimulation of the developing brain. Even if success rates are established in adults, this is no guarantee that similar results will be achieved in paediatric patients, or vice versa. If direct stimulation of the brain does not offer relief in case of adults, it might still be beneficial in the developing brain due to its greater plasticity or because protective mechanisms are triggered following stimulation.

Woopen and colleagues (2013) present the following three situations as ethically justified: performing investigational DBS in child patients for severe, treatment-refractory diseases that (a) only occur in children (e.g., some types of dystonia), (b) show severe symptoms in children such that intervention is needed to prevent serious harm (e.g., life-threatening status dystonicus), and (c) are more sensitive to successful outcomes if treated early. In sum, for certain disorders and in specific cases, investigational paediatric DBS can be ethically defensible even if success rates in adults have not been established. This is in line with article 37 of the Declaration of Helsinki (WMA 2013): “where proven interventions do not exist or other known interventions have been ineffective, the physician, after seeking expert advice, with informed consent from the patient or a legally authorized representative may use an unproven intervention if in the physician’s judgment it offers hope of saving life, re-establishing health or alleviating suffering”. Of course, the vulnerable situation of child patients obliges investigators and medical staff to be especially cautious when potentially enrolling child patients with conditions that may spontaneously remit by later adulthood.

6.5 Conclusion

If we consider the tremendous suffering that may be involved in certain severe, treatment-refractory paediatric conditions, investigational paediatric DBS can be ethically defensible even in cases where successful outcomes have not been established in adults. The starting point in case of therapeutic research with children capable of giving their assent/dissent should always be a scientifically sound hypothesis of benefit versus risk of damages and burdens. General knowledge of the side effects and possible adverse effects of DBS have to be weighed against the severity of suffering and the efficacy (or absence) of other treatment options. When the medical research team considers a child for enrolment in a paediatric DBS case series or therapeutic trial, the medical team has a professional and moral responsibility to protect and further the child's interests. The parents have a moral responsibility to decide in their child's best interests. In all cases of paediatric DBS, a shared decision-making process is therefore strongly recommended to adequately protect children's rights, to secure the best possible care and support during the treatment process, to preserve family intimacy and to stimulate the developing autonomy of the paediatric patient. A dual consent procedure is needed before the intervention can take place. Before any intervention can proceed, parents should give their consent and the child her assent. In line with the Declaration of Helsinki and the Regulations proposed by the European Parliament and the Council of Europe, children should have a veto right to refuse participation in investigational treatments, and care-giver burden should never be the main reason to initiate an invasive interventional treatment such as direct stimulation of the brain via electrode implants. Although the risk of harm due to unnecessary interventions is small in case of diseases with a clear prognosis and likelihood of irreversible damage, a dual consent procedure remains preferable because it carries greater benefits for both the parents and the children. In case of diseases such as TS that may spontaneously remit in young adulthood and for which disease prognosis is unclear, the risk of harm due to unnecessary interventions is much greater and children's right to veto should be upheld. In conclusion, in the context of direct electrical stimulation of the developing brain for neurological and psychiatric disorders, parents have the moral responsibility to take their children's developing autonomy seriously, and to respect the child's perspective.

References

Air, E.L., J.L. Ostrem, T.D. Sauger, and P.A. Starr. 2011. Deep brain stimulation in children: Experience and technical pearls. *Journal of Neurosurgery: Pediatrics* 8(6): 566–574.

Alderson, P. 1992. In the genes or in the stars? Children's competence to consent. *Journal of Medical Ethics* 18: 119–124.

DiFrancesco, M.F., C.H. Halpern, H.H. Hurtig, G.H. Baltuch, and G.G. Heuer. 2012. Paediatric indications for DBS. *Child's Nervous System* 28: 1701–1714.

Fins, J.J., H.S. Mayberg, B. Nuttin, C.S. Kubu, T. Galert, V. Sturm, K. Stoppenbrink, R. Merkel, and T.E. Schlaepfer. 2011. Misuse of the FDA's humanitarian device exemption in deep brain stimulation for obsessive-compulsive disorder. *Health Affairs* 30(2): 302–311.

Focquaert, F. 2011. Paediatric deep brain stimulation: A cautionary approach. *Frontiers in Integrative Neuroscience* 5: 1–2.

Focquaert, F. 2013. Paediatric deep brain stimulation: Parental authority versus shared decision-making. *Neuroethics* 6: 447–455.

Gilbert, R., C.S. Widom, K. Browne, D. Fergusson, E. Webb, and S. Janson. 2009. Burden and consequences of child maltreatment in high-income countries. *The Lancet* 373: 68–81.

Hagger, L. 2009. *The child as vulnerable patient. Protection and empowerment*. Farnham: Ashgate.

Hart, C., and R. Chesson. 1998. Children as consumers. *British Medical Journal* 316: 1600–1603.

Kaminska, M., D.E. Lumsden, K. Ashkan, I. Malik, R. Selway, and J.P. Lin. 2012. Rechargeable deep brain stimulators in the management of paediatric dystonia: Well tolerated with a low complication rate. *Stereotactic and Functional Neurosurgery* 90: 233–239.

Leentjens, A.F.G., V. Visser-Vandewalle, T. Temel, and F.R.J. Verhey. 2004. Manipulation of mental competence: An ethical problem in a case of electrical stimulation of the subthalamic nucleus for severe Parkinson's disease. *Nederlands Tijdschrift voor Geneeskunde* 148: 1394–1398.

Regulation EU no. 236/2014 201. http://eur-lex.europa.eu/legal-content/EN/TXT/?uri=uriserv:OJ.L_.2014.158.01.0001.01.ENG. Downloaded on 29 June 2015.

Ross, L.F. 2004. Informed consent in paediatric research. *Cambridge Quarterly of Healthcare Ethics* 13: 346–358.

Schechtman, M. 2010. Philosophical reflections on narrative and deep brain stimulation. *The Journal of Clinical Ethics* 21(2): 133–139.

Schüpbach, M., M. Gargiulo, M.L. Welter, L. Mallet, C. Béhar, J.L. Houeto, D. Maltête, V. Mesnage, and Y. Agid. 2006. Neurosurgery in Parkinson disease. A distressed mind in a repaired body? *Neurology* 66: 1811–1816.

Starr, P.A., P.S. Larson, J. Ostrem, A. Glass, W.J. Marks, M. Volz, R. Taylor, S. Heath, and J. Grace. 2015. FAQ for patients: Deep brain stimulation for dystonia. Downloaded on 28 May 2015. http://neurosurgery.ucsf.edu/index.php/movement_disorders_dystonia.html.

Unterrainer, M., and F.S. Oduncu. 2015. The ethics of deep brain stimulation (DBS). *Medicine, Health Care and Philosophy* 18(4): 475–485.

W 2004-05-07/32: www.ejustice.just.fgov.be/cgi_loi/change_lg.pl?language=nl&la=N&table_name=wet&cn=2004050732. Downloaded on 31 May 2015.

Woopen, C., K.A.M. Pauls, A. Koy, E. Moro, and L. Timmerman. 2013. Early application of deep brain stimulation: Clinical and ethical aspects. *Progress in Neurobiology* 110: 74–88.

World Medial Association. 2013. *Declaration of Helsinki. Ethical principles for medical research involving human subjects*. Guildford: Canary Publications.

Chapter 7
Neurological Diversity and Epigenetic Influences in Utero. An Ethical Investigation of Maternal Responsibility Towards the Future Child

Kristien Hens

Abstract Epigenetics is a discipline that aims to demonstrate how environmental factors influence organisms on a molecular level by explaining how these factors can affect the expression of genes. Epigenetic changes are assumed to be heritable and reversible, thus challenging the *central dogma of genetics*. This dogma states that genes are transcribed and translated unidirectionally to proteins and the genetic code cannot be changed through environmental influences. In addition, in neurology it is believed that epigenetics partly explains the development of neurological conditions and plays an important role in synaptic plasticity. As many epigenetic changes happen in utero, maternal behaviour may affect brain development. This raises questions regarding the responsibility of the pregnant woman. In this chapter, I investigate first whether and how epigenetics complicates the question of maternal responsibility, if even behaviour of years before conception may have consequences for the wellbeing of the future child. To what extent is it an individual responsibility or a collective one? Second, by using the examples of autism and high intelligence ("giftedness"), I investigate the distinction between prevention and enhancement, and demonstrate the relevance for this discussion of classifying neurological difference as an identity rather than an affliction. By analysing different concepts and arguments, I highlight the need for a new framework for maternal, parental and collective responsibility that is applicable to the context of neurology, epigenetics and beyond.

Keywords Epigenetics • Autism • Neurodiversity • Identity • Prevention • Intelligence • Enhancement • Pregnancy • Environment • Genetics

K. Hens (✉)
Department of Philosophy, University of Antwerp,
Stadscampus, Grote Kauwenberg 18, 2000 Antwerp, Belgium
e-mail: kristien.hens@uantwerpen.be

K. Hens et al. (eds.), *Parental Responsibility in the Context of Neuroscience and Genetics*, International Library of Ethics, Law, and the New Medicine 69,
DOI 10.1007/978-3-319-42834-5_7

7.1 Introduction

Epigenetics is heralded as a paradigm shift in genetics (Jablonka and Lamb 2005; Griffiths and Stolz 2013). Although it has been known for some time that both genetics and the environment contribute to the development of the phenotype, epigenetics now offers an explanation of how environmental factors influence organisms on a molecular level. This generates some challenges for bioethics and for our thinking about responsibility and identity. In this chapter I first explain the basic mechanisms of epigenetics and how these affect gene expression. I then demonstrate that these mechanisms are also relevant for the neurological development of the child. As many of these influences happen in utero, I show the way in which this raises new questions regarding the responsibility of the pregnant woman, but also about parental and collective responsibility in general.

The *central dogma of genetics* describes a unidirectional way in which genes define characteristics. In short, the genetic code is transcribed into RNA, which in turn is translated into proteins. However, already at the beginning of genetics as a science it was known that this model cannot explain certain phenomena. For example, cells that basically share the same DNA in their nucleus have different functions depending on the tissue in which they are located. Even monozygotic twins, who share the same DNA, are differently susceptible to disease. Epigenetic mechanisms can explain how genes can be switched on and off, both in the context of cell differentiation and through environmental influences. Research is currently undertaken to determine how this actually happens, and only the tip of the iceberg is known at the moment (Jablonka and Lamb 2005). One important mechanism is *methylation*: methyl marks are added on top of the genetic code. This prevents certain areas of the genetic code from being read by the transcription factors, hence the expression of that part of the DNA is prevented. Changed methylation patterns have been demonstrated, for instance when comparing sperm from the same man at different points in time, which shows that such influences happen throughout the lifespan of an organism (Jenkins et al. 2014). Another mechanism is called *histone modification*. Histones are the proteins around which DNA is wrapped. Tightening or loosening this "wrap" of DNA may also alter the accessibility of the genes located there. These mechanisms are influenced by environmental factors. Methylation and histone modification happen in the nucleus, at the moment of transcription, but certain mechanisms can also interfere with RNA translation in the cytoplasm of the cell, thus changing the way RNA is translated into proteins.

The interesting part here is that epigenetic changes may be *heritable*. A study using mice has demonstrated that if a female rat or mouse is fed a high-fat diet and becomes obese her offspring may become more vulnerable to becoming obese as well. In turn the propensity to become obese could then be passed on to several more generations, regardless of the diet these grandchildren are fed (Duhl et al. 1994; Wolff et al. 1998). Moreover, research has suggested that exposure of mice to high levels of social stress can make the offspring of those animals more susceptible to stress for several generations (Matthews and Phillips 2012).

In humans, Pembrey et al. have demonstrated that paternal grandsons of Swedish men who were exposed during preadolescence to famine in the nineteenth century were less likely to die of cardiovascular disease than those of grandfathers raised in an environment where food access was not restricted (Kaati et al. 2007). Moreover, recent studies in epigenetics show that environmental factors beyond the control of individuals may affect also future generations that are themselves no longer exposed (Thayer and Kuzawa 2011). Another novel aspect of epigenetics is that these changes are potentially *reversible*, making epigenetic changes that influence health much more readily curable than, for example, genetic mutations. In cancer, where epigenetics has been demonstrated to be one of the causes, promising research has been performed to remove methyl marks and to slow down the cancer (Falahi et al. 2014, 2015).

In the field of neurology, it is believed that epigenetics can explain the development of neurological conditions and that it plays an important role in synaptic plasticity and memory formation (Archer et al. 2011, 2012; Giovanoli et al. 2013; Zahir and Brown 2011; Levenson and Sweatt 2005; Tsankova et al. 2007; Petronis et al. 2000). Moreover, many of these epigenetic changes happen in utero and are hence susceptible to environmental factors to which the pregnant woman is exposed. For example, a study seeking the explanation for differences in IQ between monozygotic twins has found that this may be due to differences in DNA methylation and hence gene expression (Yu et al. 2012). In the same line, researchers found that choline intake during pregnancy may affect how fetal brains develop certain regions associated with memory. In the popular press this was reported under the headline that eating bacon and eggs during early pregnancy may make your future child smarter (Mehedint et al. 2010). That the chance that a child develops Attention Deficit Hyperactivity Disorder (ADHD) is increased when her mother was under stress during pregnancy has also been known for quite a while (Grizenko et al. 2012). Now it has been suggested that such stress indeed changes the methylation patterns on the DNA of the children born under stress (Grizenko et al. 2012; Monk et al. 2012).

In the context of autism, there is a gap in the understanding of how the phenotype develops. Autism is linked with mutations or changes in certain areas in the genome, but not all subjects who have changes in these areas develop autism (Schanen 2006). It is believed that epigenetics will be able to provide part of the explanation as to why certain people develop autism whereas others do not. Such epigenetic changes are linked to behavioural and environmental factors in both men and women. For example, autism has been related to older paternal age, possibly due to epigenetic changes that accumulate during the lifespan in primordial germ cells (Hamlyn et al. 2013; Ladd-Acosta et al. 2014). It has also been connected to epigenetic influences during pregnancy. The intake of folic acid, which is advised for women to reduce the chance of having a child with spina bifida, may also prevent the development of autism in future offspring (Suren et al. 2013).

7.2 The Ethics of Maternal Responsibility

Questions regarding responsibility typically involve an analysis of different aspects: *Who* is responsible for *whom* with regard to *what*? *That* parents have obligations towards their children has been defended on different grounds (Archard 2010; Archard and Benatar 2011). However, the extent to which such obligations entail specific responsibilities towards future children remains under discussion. The responsibilities of pregnant women towards their future children have been the object of much discussion over recent decades, without leading to a definite conclusion. At least it is assumed that if a woman is pregnant with a wanted child, or with a child to which she intends to give birth, this generates a special duty to behave in an appropriate way (Murray 1991). An appeal to the principle of *non-maleficence* ('do not harm your child') is assumed by many as reasonable and as applicable to the situation of the pregnant woman (Murray 1996). It seems straightforward that a woman should not smoke or drink alcohol while pregnant with a child that she intends to carry to term. Specifically in the context of substance abuse, Dondorp and de Wert argue in this volume that such abuse may even warrant coercion to prevent pregnant women's access to drugs in certain cases. But other authors have stated that women are not only mothers or pregnant vessels, and such restrictions may lack respect of individual choice (Murray 1991; Smajdor 2007). Moreover, measurements to monitor and punish behaviour of pregnant women may presuppose a too individualistic approach to responsibility: responsibility towards children presupposes at least an enabling society, and the balance between individual and collective responsibility is not at all settled.

The science of epigenetics may complicate the question of responsibility even more. Should or can we even react on the basis of that knowledge? Its implications may be beyond what people can reasonably be held responsible for. Indeed, the question about responsibility may not only be a question of 'how' but also of 'whether' and 'why'. How should one act in light of the knowledge that what one eats or is exposed to may affect the health of generations of future children? How to manage responsibilities in the context of the knowledge that a smoking 16 year old boy may—perhaps drastically—increase the chance that his future offspring develops asthma (Hens 2017)? The impact on the fetus of stress during pregnancy raises the question of how society can provide a less stressful environment to women without reducing their chances of decent employment. Moreover, the intergenerational dimension of epigenetics may inflate responsibilities to unmanageable proportions: what a pregnant woman is exposed to may influence the germ cells of the fetus she is carrying, and thus affect the health of the children of that fetus, her grandchildren. Living in polluted areas may induce epigenetic changes and affect offspring, even after families have moved to non-polluted regions. Who is responsible for someone's living in a polluted region (Thayer and Kuzawa 2011)? The responsibility could be assigned to either the person herself or those who cause the pollution, or could be conceived as a shared responsibility. We may have a collective responsibility towards all future generations, but how we must exercise that responsibility and how far this individual responsibility extends is unresolved.

The possibility of the *reversibility* of epigenetic changes may also have impact on this responsibility. If it is possible to reverse the bad influence of smoking through medication, this may relieve a pregnant woman of the duty not to smoke. Conversely, some may consider the fact that the negative outcome of smoking during pregnancy can be avoided by quitting as a sufficient argument not to spend time and money on the research into this reversibility. For example, we can tolerate that pregnant women smoke during pregnancy provided that a pill is available that can undo the bad influence of the smoking on the child. Or we could argue that smoking is behaviour under the control of an individual, and as such, rather than trying to spend money in finding a fix for the effects of smoking, policy makers should urge or even coerce people not to smoke during pregnancy. The reversibility of epigenetic changes may also function as an excuse for employers not to provide a less stressful environment for their pregnant employees, because the potential harmful effects may be reversed later.

It is clear that the science of epigenetics raises many new questions regarding responsibility. These need to be addressed before the flood of new scientific findings and inflated reports in the popular media may lead to defeatism or a laissez-faire attitude: people may feel overwhelmed by responsibilities that they cannot fulfil. Or it may lead, as demonstrated by the fictional example that Güell gives elsewhere in this volume, to exaggerated efforts to avoid all possible negative influences. Moreover, a proper reflection on the balance between individual and collective responsibilities with regard to environmental effects and personal choices needs to guide policy makers towards taking proper decisions in the light of these new findings (Juengst et al. 2014).

7.3 Epigenetics and Neurological Difference

I shall now use the example of epigenetics in the context of *neurological development in utero* to investigate aspects of responsibility more closely. I use the examples of autism and extreme intelligence to challenge some of the assumptions we may have regarding responsibility towards future offspring: in particular assumptions regarding the duty to prevent harm (*prevention*) and the duty to improve on the outcome (*enhancement*). I discuss whether pregnant women may be said to have a duty to avoid neurological conditions such as autism in their future children. Is this really an example of '*avoiding harm*'? Moreover, we can ask ourselves, do pregnant women have a duty to make their children as intelligent as possible?

7.3.1 Autism

Autism is a condition diagnosed via observation of the behaviour of the individual. Receiving a diagnosis of autism usually means that the individual exhibits stereotypical behaviour or interests and is challenged in communication and in social interaction, although the severity of symptoms may vary. Autism is thought of as a spectrum disorder, ranging from high-functioning individuals, sometimes with superior cognitive intelligence, to individuals who are severely intellectually challenged. The history of autism is laden with discussions about responsibility. The now discredited Refrigerator Theory, dating from the 1950s, stating that the origin of autistic behaviour of children lies in cold behaviour of, especially, the mother, has harmed and stigmatized parents and especially mothers, who were made to experience guilt and blame (Bettelheim 1972; Kanner 1949; Verhoeff 2013).

During recent decades, more and more genetic variants have been discovered that are linked with autism. The discovery of these genetic links has given some parents of autistic children a sense of relief, as this means that they are not responsible for the condition after all, although the knowledge that the contributing genetic mutation originated in the family of one parent may still induce feelings of guilt in the carrier of that mutation (Trottier et al. 2013). The current understanding that there are probably epigenetic components in the development of autism may complicate the genetic view (Schanen 2006). Indeed, the knowledge that environmental influences and lifestyle choices may also be responsible for the development of autism may suggest that (prospective) parents, and especially pregnant women, have a responsibility to avoid it by making the right choices (such as selecting the right nutrients during pregnancy). Hence, the implication is that they should, by avoiding autism, *avoid harm* to future offspring. This already presupposes that autism is a condition that should be prevented, in the same way as one should prevent spina bifida by taking folic acid or fetal alcohol syndrome by abstaining from alcohol intake during pregnancy. This is a view that is problematic for many, for reasons that we will look at in the following paragraphs.

7.3.2 Neurodiversity and the Pathologisation of a Normal Variation

Members of the neurodiversity movement, which originally represented autistic people but now also covers other neurological conditions such as ADHD, have suggested that a mere 'disorder' approach to such conditions is unsatisfactory. Many of its members claim that autism is not a disease, and hence need not be cured, and that it is associated with benefits that are often overlooked by 'neurotypical' others (Fenton and Krahn 2007; Ortega and Choudhury 2011; Hacking 2009; Glannon 2007). Instead, they claim that the difficulties associated with autism are not intrinsically related to the condition, but are due to a lack of acceptance from society. They mirror the views of members of the disability rights movement who claim that disabilities are primarily defined by social context and that society has a responsibility to accept and accommodate difference. The International Declaration on the Rights of Persons with Disabilities (IDRPD) has explicitly embraced this *social model of disability*, which is distinct from the so-called *medical model of disability*. The latter views disability as a condition intrinsic to the individual, which limits her quality of life and which should be cured by medical interventions.

In her chapter in this volume, Bosman argues that DSM-based diagnoses of disorders such as autism are not scientific but the result of value judgments and cultural norms. Following Canguilhem she claims that pathology is not localised *in* the individual (or *in* the environment for that matter) but that it is the diminished possibilities of the organism in that particular environment to conform to existing norms. Indeed, in the case of autism, the aspect of dysfunctioning seems to be a prerequisite for being labelled. This would make the issue of whether one is diagnosed with a pathology or not at least partly contingent on how supportive the environment is.

Members of the neurodiversity movement stress the fact that they speak a different language which neurotypicals cannot or refuse to speak. Autism also has positive aspects, which neurotypicals lack. Those with autism are often visual thinkers with a great eye for detail, and their metonymical experience of the world may shed a new light on reality that is inaccessible to most other people. But autism is a spectrum condition, and those on the far end of the spectrum may equally well be unable to benefit from the advantages that some high-functioning individuals have. Moreover, aspects of autism may hamper some individuals' functioning regardless of the support of society. For example, a person may still suffer from sensitivity to noises, even if everything possible is done to prevent most of those noises, as she will have to confront some level of noise at some point in life. Plus, a low-functioning autistic child may very well develop into a high-functioning adult and a child that was functioning reasonable well may have severe challenges in adulthood (Walsh et al. 2011).

7.3.3 Autism and Identity

A recurrent theme in the neurodiversity movement is the question of *identity*. Being autistic is seen by many of the movement's members as an integral part of their identity, and therefore the discussion about a cure is irrelevant, as this would fundamentally change who they are. Epigenetic influences in utero may change brain development in such a way that the resulting child has a differently functioning brain than she would have had otherwise. Thus, such changes, leading to neurodifference, may be identity-affecting. Indeed, in the context of gene therapy, Glannon has argued that the earlier a genetic intervention to prevent or cure a physical disorder occurs in one's biological life, the more likely it would be identity-determining rather than identity-preserving (Glannon 2001). Although epigenetic influences do not actually change the genetic code itself, they define which genes are expressed and which not. If they occur in utero I believe that they may equally be considered identity-determining and the difference with actually intervening in the DNA itself is trivial. This has relevance for the question *for whom* prospective parents are responsible. If the behaviour of a pregnant woman can cause certain neurological changes in the fetus, such that given those changes it would develop into a child whose behaviour would be labelled autistic, this behaviour may cause a child to come into existence with an identity other than the one she would have had, had the woman's behaviour been different (e.g. an autistic identity versus a non-autistic one). The responsibility may no longer be towards the wellbeing of a given child, but towards choosing *the best child*. Moreover, the fact that autism is considered a spectrum disorder, with ranges from high-functioning individuals to low-functioning people with severe intellectual disability, raises the question of whether a disease or disorder can also be an identity at the same time, and whether certain identities should be avoided or not.

An analogy can be drawn with the discussion on embryo selection, which is by definition identity-affecting. In this context, certain authors have specified threshold principles, defining thresholds that an embryo should meet before being eligible for transfer to the womb. Those who propose a minimal threshold imply that only those embryos should be discarded (or not transferred) that would develop into children whose life would be filled with suffering to the degree that it would allow no enjoyment. Others define an acceptable outlook threshold or even a maximum threshold, stating that only embryos with a very good prospect should be transferred (Glover 2006; Harris 2001). The question of whether future parents should be concerned with which children there should be reflects the "external perspective" described by Herissone-Kelly in this volume. This is distinct from their responsibility towards the wellbeing of specific children ("the internal perspective"), and is not yet resolved.

7.3.4 Intelligence

Increasing intelligence is often presented as the ultimate example of enhancement. For example, Julian Savulescu has stated that, should such information be available in vitro, 'smarter' embryos (IQ >140) should be chosen, as such people are expected to flourish most (Savulescu 2001). Moreover, high intelligence and the associated heightened autonomy are often quoted as examples of enhancement in post-humanist writings in general (Glover 2006). The value that is attributed to intelligence is also high in the advice that is given to pregnant women and to parents of young children. Pregnant women are advised to eat fish as the omega-3 will help brain development and hence intelligence. Musical education for children is praised, not solely for the pleasure of playing music but often also because it is deemed to increase a child's intelligence (Hille et al. 2011). One need only enter a toy store to understand that many toys today are developed and presented, not primarily as tools for amusement, but as IQ-enhancing. There is an explosion of so-called smart games, whose function is not, so it seems, to satisfy the needs of the more cognitively oriented child, but actually to develop intelligence.[1] So it seems that intelli-

[1] For example, the company SmartGames advertises: "Have fun exercising your brain! Utilize skills ranging from spatial insight to pattern recognition, while improving your problem solving abilities and your memory".

gence is valued and that this value increments linearly with the intelligence. The smarter your kid is, the better.

The complex problems of gifted children suggest that this view may be unsatisfactory (Reis and McCoach 2000; Roedell 1986, 1984). Children with IQs over 130 are considered 'gifted' and are in many schools in many countries entitled to special care. Consensus arises more and more that this is a vulnerable group requiring special support (Roedell 1984). Many of the problems these children face are related to their *social* environment. On the social level, it can be deemed a problem that the educational system is not adapted to those who fall on either side of the Gauss curve. A system that has enough flexibility to adapt to all learning styles may solve many problems. If these children have problems with fitting in with their peer groups, or are even bullied, this could be mitigated by facilitating extracurricular contact with peers with the same interests (Peterson and Ray 2006). Educating children in general to embrace and respect otherness may solve some of the problems.

Note that all these measures take up resources. Moreover, one might argue that there are *intrinsic* risks associated with being extremely intelligent. Intelligence may also be linked with an increased awareness and autonomy. It has been suggested that being able to understand the grand picture of life and one's place in this grand picture may lead to an increased risk of existential anxiety (Webb et al. 2005). Admittedly, the link between extreme intelligence and depression or suicide has not yet been adequately researched. But to automatically assume that higher intelligence leads to higher wellbeing seems untenable as well. There seems to be considerable overlap between the behaviour of high-functioning autistic individuals and highly intelligent gifted people (Webb et al. 2005). This link between autism and high intelligence may even be genetic (Clarke et al. 2016). In her most recent book, Grandin gives the example of successful young people she met in Silicon Valley, whom she compares to youngsters she met in other areas of the US. Both groups shared the same characteristics, she said, but in the former case this was considered a gift, whereas in the latter it was considered a challenge, as these children were expected to behave like typically developing children (Grandin and Panek 2014). This raises the question of whether the distinction between what is a desirable trait and what is a disability is difficult, and perhaps impossible, to draw (Mertes and Hens 2015).

7.4 Neurological Diversity and Responsibility

An investigation of maternal responsibility inevitably leads to the question of *what* a pregnant woman is responsible for. The existing debates have focused on whether she has the responsibility to prevent harm, or to give her child the best life possible. Does she have the responsibility of making sure that her fetus is not harmed, or that it is enhanced to maximise the child's opportunities later in life? It seems straightforward that one desires the best for one's child. But whether the categories of *prevention* versus *enhancement* are really useful in the light of complex phenomena involving neurological development, such as autism spectrum disorder or high intelligence is still unresolved. One may wonder whether they are useful categories for discussing the responsibility of the pregnant woman and parental responsibilities *tout court*.

7.4.1 Two Fictional Cases: Mary and John

If we assume that harm should be prevented and wellbeing increased, we need first to be able to define what we mean by harm and wellbeing. A commonly accepted approach would be to refer to the wellbeing the individual herself reports. Consider the fictional Mary, a 25-year-old woman with a diagnosis of autism who lives with her family and spends her days in her room discussing and gaming on the internet. Mary may very well consider her own quality of life good. However, the outside world may consider Mary's life not to be representative of the good life, as she lacks the autonomy that others value. Moreover, Mary's parents have repeatedly told other parents that they regret the fact that she lacks the capabilities to live a life independent from them. They would accept a cure that would make Mary more autonomous.

Consider John, an academic who is on the verge of a breakthrough in the application of nuclear fusion that will solve many of the world's energy and ecological problems. He is an introvert who feels unconnected with his fellow human beings and has had bouts of severe depression. Is the potential good for society that John may generate sufficient reason to prefer his life over that of Mary, regardless of his own wellbeing?

An often-expressed critique from the neurodiversity movement and the disability rights movement in general is that it is not a condition itself that is problematic, but the reaction and support of society (Wilkinson 2010). The lack of wellbeing that many autistic people may experience is hence linked to a social problem rather than an intrinsic one related to a medical pathology. Indeed, in her book *Disability Bioethics*, Scully (2008) has pinpointed a weak spot in traditional bioethical thinking, and noted why bioethics has failed to adequately take into account the perspectives of disabled people. She writes that

> Bioethicists have mostly been called upon to evaluate ethically and help regulate the new medical technologies of prenatal or preimplantation screening for impairments, and the voluntary or involuntary euthanasia of disabled people. Combined with its symbiotic relationship with biomedicine, this limited range of tasks has meant that bioethics reflexively turns to the conceptual framework of medicine for thinking about disability.

Hence she claims that many bioethicists have assumed the medical model of disability, which sees disability mainly as impairment. She states that this approach to disability issues has a strongly libertarian and utilitarian flavour, and systematically neglects disability studies as an important source of insight into real moral lives (Scully 2008). The failure of much bioethical thinking to deal with disability also becomes apparent when investigating parental and collective responsibility. Indeed, parental responsibility towards the welfare of specific children cannot be conflated with collective responsibility towards the welfare of all children. It may be the case though, that, as Hans Jonas has argued, parental responsibility is an archetype for responsibility as such, and can function as a blueprint for collective responsibility, allowing for much greater consideration of the specifics of individual children rather than of the characteristics of the population as a whole (Jonas 1984).

7.4.2 Parental Responsibility and Neurological Development

But maybe it does not matter what the *cause* of reduced wellbeing is, be it intrinsic or social. We have used the example of epigenetic influences in utero, but parents in general, prospective or actual, may have a responsibility to avoid having children who would not fit into a given society without support, even though they may not have a problem *per se*. After all, this may be an easier fix than trying to change society as a whole. Moreover, this may require (prospective) parents not only to avoid autism in their children, but also to avoid extreme intelligence, as those who are moderately intelligent may face fewer social difficulties than those who are extremely intelligent. Hence, this would set certain boundaries to the goal of enhancement, and may even redefine the content of the concept of cognitive enhancement itself, as enhancing would then imply ensuring that children conform to the normal. However, it is very difficult to define a characteristic that is good, as this will depend on what factors are included in the calculus, and whichever characteristics we choose is relative to the context.

Although common sense would dictate that the match with society is also important, to state that parents have a responsibility to make sure that only those types of people who fit within a given society are born is an unsavoury thought as well. At this point it is unclear whether, albeit unsavoury, it is a valid point. Although diversity is perhaps not much valued or accepted in present-day society, this does not mean that it is not a value that we *should* endorse. Indeed, game theory predicts that if a large number of similar 'perfect' people were to be created, it would soon be better *not* to resemble them but to possess complementary characteristics and abilities (Mertes and Hens 2015). In fact, many policy decisions seem to suggest that diversity is considered a value. For example, in Belgium, since 2014, the 'M-decreet'[2] is put into place. This is a set of legal measurements aimed at a better integration of children with certain challenges in the standard curriculum. These measurements intend to stimulate, by valuing inclusion of children with special needs, more diversity in the classroom, rather than automatically sending these children to special-

[2] "M-Decreet", stands for "Maatregelen voor leerlingen met specifieke onderwijsbehoeften", which translates to "measurements for pupils with special educational needs".

education schools. Moreover, it is unclear what part of society one should fit into or feel good in. A child with autistic features unable to cope in a classroom of 25 pupils may function very well in adult life where she has the autonomy to choose her own environment and co-workers.

The balance between individual wellbeing and what is good for the collective is also unclear. On the one hand, certain people may function very well in a society provided that enough accommodation is provided to address their needs. Do economic aspects form part of the calculus? On the other hand, a person with extreme intelligence may have much to offer with regard to the progress of science, but may be suffering from bouts of existential loneliness. This may raise the question of whether we, and parents in particular, have a responsibility to create persons with perhaps lesser wellbeing, who may as a result of their genius solve the world's energy crisis through solving the puzzle of nuclear fusion. Whose wellbeing should prevail here? The particularities of being a parent, which involve duties of care, may in effect rule out that such considerations should be part of parental responsibilities.

7.5 Conclusion

Analysing the responsibility of a pregnant woman towards her fetus involves an analysis of at least the issue of *who* is responsible for *whom* with regard to *what*. I have used the example of epigenetics and neurology to demonstrate that these aspects are not explored in sufficient depth to provide an answer. With regard to *who*, it is clear that the balance of individual and collective responsibility is not decided. Discussions about substance abuse and smoking in pregnancy often stress the individual responsibility of the pregnant woman, leading ultimately to a very punitive and stigmatizing approach. However, commentators have already pointed out that such an approach is unsatisfactory, and that a focus on individual responsibility neglects the social context.

In the context of neurology the question of *for whom* is important as well. If a neurological difference such as autism is an identity, as the proponents of the neurodiversity movement state, a pregnant woman who through environmental exposure effectively changes the neurological status of her fetus so that it will not develop into a child with ASD would effectively change the identity of the fetus. This would mean that she would not be doing something for the wellbeing of a specific child, but rather, by avoiding or reducing autism, creating one child rather than another. I claim this distinction is relevant and in need of further exploration. The relation of a pregnant woman, and in general of (prospective) parents, with their child is a particularistic one, involving duties of care towards that specific child. Whether these duties also include a more consequentialist approach, implying that the calculus involves deciding to have one child rather than another, is as yet unresolved.

Acknowledgement Kristien Hens is supported by a Belgian Federal Science Policy Office (BELSPO) "Back to Belgium" grant.

References

Archard, D. 2010. The obligations and responsibilities of parenthood. In *Procreation and parenthood: The ethics of bearing and rearing children*, ed. D. Archard and D. Benatar. Oxford: Oxford University Press.

Archard, D., and D. Benatar. 2011. *Procreation and parenthood: The ethics of bearing and rearing children*. Oxford: Oxford University Press.

Archer, T., M. Oscar-Berman, and Blum, K. 2011. Epigenetics in developmental disorder: ADHD and Endophenotypes. *Journal of Genetic Syndrome & Gene Therapy* 2: 1–7.

Archer, T., M. Oscar-Berman, K. Blum, and M. Gold. 2012. Neurogenetics and epigenetics in impulsive behaviour: Impact on reward circuitry. *Journal of Genetic Syndrome & Gene Therapy* 3: 1000115.

Bettelheim, B. 1972. *The empty fortress: Infantile autism and the birth of the self* (New edition). New York: Free Press.

Clarke, T.K., M. K. Lupton, A.M. Fernandez-Pujals, J. Starr, G. Davies, S. Cox, and A.M. McIntosh. 2016. Common polygenic risk for autism spectrum disorder (ASD) is associated with cognitive ability in the general population. *Molecular Psychiatry* 21(3): 419–425.

Duhl, D.M.J., H. Vrieling, K.A. Miller, G.L. Wolff, and G.S. Barsh. 1994. Neomorphic agouti mutations in obese yellow mice. *Nature Genetics* 8(1): 59–65.

Falahi, F., M. van Kruchten, N. Martinet, G.A.P. Hospers, and M.G. Rots. 2014. Current and upcoming approaches to exploit the reversibility of epigenetic mutations in breast cancer. *Breast Cancer Research: BCR* 16(4): 412.

Falahi, F., A. Sgro, and P. Blancafort. 2015. Epigenome engineering in cancer: Fairytale or a realistic path to the clinic? *Frontiers in Oncology* 5: 22.

Fenton, A., and T. Krahn. 2007. Autism, neurodiversity and equality beyond the "Normal". *Journal of Ethics in Mental Health* 2: 1–6.

Giovanoli, S., H. Engler, A. Engler, J. Richetto, M. Voget, R. Willi, and U. Meyer. 2013. Stress in puberty unmasks latent neuropathological consequences of prenatal immune activation in mice. *Science* 339(6123): 1095–1099.

Glannon, W. 2001. *Genes and future people. Philosophical issues in human genetics*. Boulder: Westview Press.

Glannon, W. 2007. Neurodiversity. *Journal of Ethics in Mental Health*. 2: 1–6.

Glover, J. 2006. *Choosing children: Genes, disability, and design: genes, disability, and design*. Oxford: Oxford University Press.

Grandin, T., and Panek, R. 2014. *Het autistische brein / druk 1: Voorbij het spectrum denken*. Trans: W. Scheen (1. ed.). Amsterdam: Nieuwezijds B.V.

Griffiths, P., and K. Stolz. 2013. *Genetics and philosophy. An introduction*. Cambridge: Cambridge University Press.

Grizenko, N., M.E. Fortier, C. Zadorozny, G. Thakur, N. Schmitz, R. Duval, and R. Joober. 2012. Maternal stress during pregnancy, ADHD symptomatology in children and genotype: Gene-environment interaction. *Journal of the Canadian Academy of Child and Adolescent Psychiatry = Journal de l'Academie Canadienne de Psychiatrie de L'enfant et de L'adolescent* 21: 9–15.

Hacking, I. 2009. Humans, Aliens & Autism. *Daedalus* 138: 44–59.

Hamlyn, J., M. Duhig, J. McGrath, and J. Scott. 2013. Modifiable risk factors for schizophrenia and autism — Shared risk factors impacting on brain development. *Neurobiology of Disease* 53: 3–9.

Harris, J. 2001. One principle and three fallacies of disability studies. *Journal of Medical Ethics* 27(6): 383–387.
Hens, K. 2017. The ethics of postponed fatherhood. *International Journal of Feminist Approaches to Bioethics*. 10(1) (in press).
Hille, K., K. Gust, U. Bitz, and T. Kammer. 2011. Associations between music education, intelligence, and spelling ability in elementary school. *Advances in Cognitive Psychology* 7: 1–6.
Jablonka, E., and M.J. Lamb. 2005. *Evolution in four dimensions. Genetic, epigenetic, behavioral and symbolic variation in the history of life*. Cambridge, MA: MIT Press.
Jenkins, T.G., K.I. Aston, C. Pflueger, B.R. Cairns, and D.T. Carrell. 2014. Age-associated sperm DNA methylation alterations: Possible implications in offspring disease susceptibility. *PLoS Genetics* 10(7): e1004458.
Jonas, H. 1984. *Das Prinzip Verantwortung. Versuch einer Ethik für die technologische Zivilisation*. Frankfurt/M: Suhrkamp.
Juengst, E.T., J.R. Fishman, M.L. McGowan, and R.A. Settersten Jr. 2014. Serving epigenetics before its time. *Trends in Genetics* 30(10): 427–429.
Kaati, G., L.O. Bygren, M. Pembrey, and M. Sjöström. 2007. Transgenerational response to nutrition, early life circumstances and longevity. *European Journal of Human Genetics: EJHG* 15(7): 784–790.
Kanner, L. 1949. Problems of nosology and psychodynamics of early infantile autism. *The American Journal of Orthopsychiatry* 19(3): 416–426.
Ladd-Acosta, C., K.D. Hansen, E. Briem, M.D. Fallin, W.E. Kaufmann, and A.P. Feinberg. 2014. Common DNA methylation alterations in multiple brain regions in autism. *Molecular Psychiatry* 19(8): 862–871.
Levenson, J.M., and J.D. Sweatt. 2005. Epigenetic mechanisms in memory formation. *Nature Reviews Neuroscience* 6(2): 108–118.
Matthews, S.G., and D.I. Phillips. 2012. Transgenerational inheritance of stress pathology. *Experimental Neurology* 233(1): 95–101.
Mehedint, M.G., M.D. Niculescu, C.N. Craciunescu, and S.H. Zeisel. 2010. Choline deficiency alters global histone methylation and epigenetic marking at the Re1 site of the calbindin 1 gene. *The FASEB Journal* 24(1): 184–195.
Mertes, H., and K. Hens. 2015. The diversity of genetic perfection. *The American Journal of Bioethics* 15(6): 34–36.
Monk, C., J. Spicer, and F.A. Champagne. 2012. Linking prenatal maternal adversity to developmental outcomes in infants: The role of epigenetic pathways. *Development and Psychopathology* 24(4): 1361–1376.
Murray, T.H. 1991. Prenatal drug exposure: Ethical issues. *Drug-Exposed Infants* 1: 105–112.
Murray, T.H. 1996. *The worth of a child*. Berkeley: University of California Press.
Ortega, F., and S. Choudhury. 2011. Wired up differently: Autism, adolescence and the politics of neurological identities. *Subjectivity* 4(3): 323–345.
Peterson, J.S., and K.E. Ray. 2006. Bullying and the gifted: Victims, perpetrators, prevalence, and effects. *Gifted Child Quarterly* 50(2): 148–168.
Petronis, A., I.I. Gottesman, T.J. Crow, L.E. DeLisi, A.J. Klar, F. Macciardi, and G.R. Sutherland. 2000. Psychiatric epigenetics: A new focus for the new century. *Molecular Psychiatry* 5(4): 342–346.
Reis, S.M., and D.B. McCoach. 2000. The underachievement of gifted students: What do we know and where do we go? *Gifted Child Quarterly* 44(3): 152–170.
Roedell, W. 1984. Vulnerabilities of highly gifted children. *Roeper Review* 6(3): 127–130.
Roedell, W. 1986. Socioemotional vulnerabilities of young gifted children. In *Intellectual giftedness in young children*, ed. J. Whitmore, 17–30. New York: The Haworth Press.
Savulescu, J. 2001. Procreative beneficence: Why we should select the best children. *Bioethics* 15(5–6): 413–426.
Schanen, N.C. 2006. Epigenetics of autism spectrum disorders. *Human Molecular Genetics* 15(suppl 2): R138–R150.

Scully, J.L. 2008. *Disability bioethics: Moral bodies, moral difference*. Lanham: Rowman & Littlefield.

Smajdor, A. 2007. The moral imperative for ectogenesis. *Cambridge Quarterly of Healthcare Ethics* 16(3): 336–345.

Suren, P., C. Roth, M. Bresnahan, M. Haugen, M. Hornig, D. Hirtz, and C. Stoltenberg. 2013. Association between maternal use of folic acid supplements and risk of autism in children. *JAMA: The Journal of the American Medical Association* 309(6): 570–577.

Thayer, Z.M., and C.W. Kuzawa. 2011. Biological memories of past environments: Epigenetic pathways to health disparities. *Epigenetics: Official Journal of the DNA Methylation Society* 6(7): 798–803.

Trottier, M., W. Roberts, I. Drmic, S.W. Scherer, R. Weksberg, C. Cytrynbaum, and F.A. Miller. 2013. Parents' perspectives on participating in genetic research in autism. *Journal of Autism and Developmental Disorders* 43(3): 556–568.

Tsankova, N., W. Renthal, A. Kumar, and E.J. Nestler. 2007. Epigenetic regulation in psychiatric disorders. *Nature Reviews. Neuroscience* 8: 355–367.

Verhoeff, B. 2013. Autism in flux: A history of the concept from Leo Kanner to DSM-5. *History of Psychiatry* 24(4): 442–458.

Walsh, P., M. Elsabbagh, P. Bolton, and I. Singh. 2011. In search of biomarkers for autism: Scientific, social and ethical challenges. *Nature Reviews Neuroscience* 12(10): 603–612.

Webb, J.T., E.R. Amend, N.E. Webb, J. Goerss, P. Beljan, and F.R. Olenchak. 2005. *Misdiagnosis and dual diagnoses of gifted children and adults: ADHD, Bipolar, OCD, Asperger's, Depression, and other disorders*. Scottsdale: Great Potential Press.

Wilkinson, S. 2010. *Choosing tomorrow's children. The ethics of selective reproduction*. Oxford: Clarendon.

Wolff, G.L., R.L. Kodell, S.R. Moore, and C.A. Cooney. 1998. Maternal epigenetics and methyl supplements affect agouti gene expression in Avy/a mice. *FASEB Journal: Official Publication of the Federation of American Societies for Experimental Biology* 12(11): 949–957.

Yu, C.-C., M. Furukawa, K. Kobayashi, C. Shikishima, P.-C. Cha, J. Sese, and T. Toda. 2012. Genome-wide DNA methylation and gene expression analyses of monozygotic twins discordant for intelligence levels. *PLoS One* 7(10): e47081.

Zahir, F.R., and C.J. Brown. 2011. Epigenetic impacts on neurodevelopment: Pathophysiological mechanisms and genetic modes of action. *Pediatric Research* 69(5 Pt 2): 92R–100R.

Chapter 8
Prenatal Child Protection. Ethics of Pressure and Coercion in Prenatal Care for Addicted Pregnant Women

Wybo Dondorp and Guido de Wert

Abstract If pregnant women are unwilling or unable to assume their responsibility for the health and wellbeing of the child they are carrying, prenatal child protection becomes a societal and professional concern. As both pressure and coercion interfere with a person's right to self-determination, such measures require justification. It is not sufficient that prenatal child protection is a morally important goal. The harm to be prevented must be plausible, and the measures must fulfil the criteria of effectiveness, proportionality and subsidiarity. Coercion always also requires a legal title. The concern that coercion would only be possible in later stages of pregnancy in order not to jeopardize women's right to have an abortion, mistakenly assumes that prenatal child protection is about the fetus. However, it is about the future child.

Keywords Pressure • Coercion • Pregnancy • Addiction • Prenatal child protection • Autonomy • Paternalism • Substance abuse

8.1 Introduction[1]

In the past years, there is growing attention for the protection of children against possible threats to their health and wellbeing that may already arise prenatally. This includes the introduction of preconception and early pregnancy educational programmes aimed at providing information about lifestyle related risk factors. The

---

[1]A draft of this paper was presented by the first author at the Conference 'Parental Responsibility in the Context of Neuroscience and Genetics', Maastricht 9, December 2013. This paper is a significantly revised and extended version of Dondorp and de Wert 2015. This work was funded by The Netherlands Organisation for Health Research and Development (ZonMw), The Hague. Project: 'The use of pressure, coercion and compulsion in the care for addicted pregnant women: an ethical analysis', project number: 141010002.

W. Dondorp (✉) • G. de Wert
Department of Health, Ethics & Society, School for Public Health
and Primary Care (CAPHRI), Maastricht University, Maastricht, The Netherlands
e-mail: w.dondorp@maastrichtuniversity.nl

K. Hens et al. (eds.), *Parental Responsibility in the Context of Neuroscience and Genetics*, International Library of Ethics, Law, and the New Medicine 69,
DOI 10.1007/978-3-319-42834-5_8

presupposition in those programmes is that prospective mothers, if well-informed about those risks, will as a matter of course adapt their lifestyle and refrain from behaviour that may be harmful for their future child. Whereas most women who actively plan a pregnancy or who know that they are pregnant will indeed regard this as their responsibility and act accordingly, things are less obvious for women addicted to drugs that may interfere with their child's development, potentially leading to learning and behavioural disorders and other adverse health effects. As professional efforts to inform these women about the need to change their behaviour may not lead to the required response, the question arises under which conditions (if any) it would be acceptable for professionals providing prenatal care to these women to use pressure or coercion in order to protect the future child from harm. In this paper we will argue that although taking such measures may well be acceptable in specific cases, their justification depends on multiple conditions that must all be met. We thereby argue both against the view that in prenatal care respect for women's autonomy is the single most important principle that trumps all other considerations, including the interests of the future child, and against the opposite position that prenatal child protection measures bring their own justification, regardless of the interests of the pregnant woman.

The set-up of this paper is as follows. We start with a brief Sect. 8.2 explaining the concepts of 'pressure' and 'coercion'. We will then in Sect. 8.3 subsequently discuss the conditions for the ethical justification of the relevant measures. In a further Sect. 8.4 we argue that for the proportionality of the envisaged measures, it does make a difference whether the woman in question is or is not to be regarded as competent with regard to decisions about lifestyle or behaviour during pregnancy. As coerced hospitalization involves infringement of what in most countries is recognised as a legal right to self-determination, it not only requires ethical, but also legal justification. We will discuss this in Sect. 8.5, using recent Dutch jurisprudence as an example of how judges try to strike a balance between the interests at stake. This then leads to a further discussion Sect. 8.6 of an issue with a wider relevance also in other jurisdictions, namely whether the increasing legal scope for prenatal child-protection measures also at earlier stages of pregnancy may lead to undermining the woman's right to ask for an abortion. We will argue that this tension disappears if those measures are seen as protecting the future child rather than the fetus. The final Sect. 8.7 contains our conclusions, partly cast in the form of a practice aid.

8.2 Pressure and Coercion

The terms pressure and coercion refer to a spectrum of measures aimed at bringing a person to behave in a specific way (Schermer 2003). They range from a relatively minor to a very major interference with a person's freedom to actively determine her own life. *Pressure* involves trying to influence a person's behaviour, but without completely denying her a choice in the matter. The least invasive form of pressure is explicit directive counselling, where professionals convey the message that patients or clients should do or refrain from doing something. This is to be distinguished from non-committal and non-directive advice about the pros and cons of a specific choice. Strong pressure entails steering choices by attaching a bonus or penalty to the desired or undesired behaviour. For instance, in the Dutch city of Rotterdam, professionals use a protocol under which a request for legal supervision is made for all pregnant women found to be addicted to a wide range of drugs (heroin, cocaine, party drugs, etc.). This is done despite the fact that such requests are only granted by the courts at a later stage in pregnancy (Bijlsma et al. 2008), and only recently also prior to 24 weeks. Routinely sending in such requests (much) earlier in pregnancy works as an instrument of strong pressure: women know that this threat is hanging over their heads and that if they do not comply (with e.g. regular urine controls) the request may be granted by the court and that this may eventually lead to their child being taken away from them (Boonekamp et al. 2012). Finally, *coercion* means the complete denial of a free choice as in coerced hospitalization or coerced treatment.

8.3 Ethical Justification

As measures of pressure or coercion interfere with a person's right to self-determination, they are in need of ethical justification (Boonekamp et al. 2012). We argue that such ethical justification would involve the following conditions all being met.

8.3.1 Morally Important Goal

Firstly, the proposed measures must be aimed at a morally important goal that can in principle be recognized as such by all parties involved. This is clearly the case if the goal is to protect the future children of addicted pregnant women from harm, in situations where those women are unable or unwilling to take their own responsibility in this regard. That pregnant women have this responsibility can in most cases be construed as a form of prenatal parental responsibility, deriving from the fact that, with her partner, she is 'retrospectively' or at least 'causally' responsible for starting the chain of events that will normally lead to the birth of a child. But as Björnsson and Brülde explain in this volume, these are only two possible sources of normative responsibility. Even if the pregnancy were to result from rape, the woman still has a normative responsibility arising from what they refer to as 'capacity': no one else but the pregnant woman is in as good a position to provide a safe developmental environment for the future child. Clearly, this presumes that the pregnancy will indeed lead to the birth of a child; if the woman decides to have an abortion there will be no birth, and therefore no child who can already be harmed prenatally. Therefore, we hold that unless they intend to have an abortion at a stage of pregnancy where termination is still legally possible, women who know they are pregnant have a normative responsibility to as much as reasonably possible to protect the future child from avoidable harm. In situations where they are unable or unwilling to assume this responsibility, prenatal child protection becomes a morally important goal for society, more particular for professionals involved in the care of the women in question. Note that 'prenatal child protection' as presented here is a matter of protecting the *future child*, rather than the fetus. This distinction is important given the intractable nature of the debate about the moral status of the fetus. Only as aimed at protecting the future child can prenatal child protection be understood as a goal the moral importance of which can in principle be recognized by all parties involved.

8.3.2 Harm Plausibility

One may think here of protecting the future child against: a) health damage already occurring in utero as a result of fetal exposure to toxic substances (prenatal harm), b) health damage occurring at birth as a result of avoidance or refusal of necessary obstetric care (perinatal harm), c) immediate and possible long-term harm if withdrawal symptoms are left untreated (neonatal harm), and d) damage to its health and welfare as a result of being raised in unsafe circumstances (postnatal harm). However, it must at least be plausible that the situation or outcome to be avoided would otherwise occur. For instance, with regard to such measures being aimed at avoiding that the child will be harmed as a result of untreated Neonatal Abstinence Syndrome (NAS) after being exposed to prenatal heroine, methadone or other drugs associated with withdrawal symptoms (tremors, irritability, excessive crying, diarrhoea), there is sufficient evidence from the literature that this is indeed a realistic concern (Kocherlakota 2014). If children are born to women who avoid prenatal care, NAS may remain untreated which leads to immediate harm in the form of unnecessary suffering and may in severe cases even cause the child's death (Schaefer et al. 2007). With regard to measures aimed at avoiding 'prenatal harm', one would need to show what evidence there is for the impact of exposure to the relevant substance (or substances) on normal fetal development, taking account of the dose and timing of exposure. Unfortunately, due to methodological problems that cannot easily be avoided when conducting research in this area (retrospective design, self-reported intake, confounding factors including poly drug use), the body of evidence showing such effects is generally not very strong. As will be discussed below, this is highly relevant in view of the proportionality of specific measures. Here, however, it suffices to observe that the uterus is not a safe haven and that harmful effects of specific types of drug abuse during pregnancy are indeed plausible.

8.3.3 *Effective Measures*

Thirdly, the proposed measure must effectively lead or contribute to the prevention of the harm in question. For instance, coerced hospitalization in the final stages of pregnancy may be an effective measure to avoid the harmful effects of untreated NAS, but if intended to avoid fetal exposure, in order to be effective, such measures must be taken much earlier, given that (depending on the specific type of drug) most of the damage will occur in early pregnancy. Along the same lines, it is also important to be aware of possible counterproductive effects, as when increased levels of pressure may chase women away from health providers. Where specific measures are introduced, it is important to evaluate their effectiveness in due time.

8.3.4 *Proportionality*

Fourthly, measures must be proportional, meaning more specifically that the level of interference in the life and freedom of the woman must be proportional to the probability and seriousness of the harm to be avoided. Take the case of cocaine abuse during pregnancy. Whereas a recent literature review of the health effects of cocaine abuse during pregnancy lists several associations both with adverse pregnancy outcomes and poor postnatal neurological development, the authors also point to several confounding factors related to 'the context of other social and environmental risk factors found in disadvantaged populations' that together make it difficult to draw firm conclusions about the precise harmfulness of prenatal exposure to maternal cocaine abuse (Cressman et al. 2014). According to recent studies, adverse health effects connected to prenatal cocaine abuse seem less serious than previously thought (Buckingham-Howes et al. 2013). Therefore, it must be concluded that coercive (and punitive) measures as were taken against crack using pregnant women in the USA in the eighties of the last century (crack being a form of cocaine that can be smoked) were disproportional. The children of these women (called 'crack babies' at the time) have developed into adults that are generally doing well (Okie 2009).

Another example is prenatal alcohol abuse. Women who drink heavily, or who indulge in incidental binge drinking, put the child at a high risk of the serious harms connected to fetal alcohol spectrum disorders (FASD) or the even more serious fetal alcohol syndrome (FAS), involving specific facial features, growth retardation and neurodevelopmental abnormalities (Health Council of the Netherlands 2005). As this makes alcohol abuse during pregnancy more dangerous for the future child than cocaine, would coerced hospitalization in order to keep pregnant women from dangerous levels of drinking be proportional? Apart from the challenge of precisely defining those levels (Robinson et al. 2010), a complicating factor is that most of the damage to the developing fetus occurs already at the early stages of pregnancy, meaning that coerced hospitalization at later stages will be of limited effectiveness.

In order to avoid this, it would be necessary to keep women hospitalized against their will for almost the full duration of the pregnancy, which clearly amounts to a much more serious infringement of their right to self-determination than doing so for a more limited period of time at a later stage of pregnancy (e.g. to avoid unsupported birth giving). And even if, given the strong evidence that heavy drinking puts the child at a high risk of serious harm, such long term hospitalization would still seem proportional, this option may not be feasible for practical reasons (the initial damage being done before even the woman herself knows she is pregnant) and legal barriers to hospitalization of pregnant women in the interest of the future child.

With regard to light drinking (1–2 units per week), there is no evidence of an adverse effect on fetal development or on the health and wellbeing of the child (Bakker et al. 2010; Kelly et al. 2013). This means that not only coercion to force women to complete abstinence would be disproportional, but also all forms of pressure, including directive counselling: if there is no harm in drinking one or two glasses a week, then clearly women should be allowed to make their own choices in this regard? Health authorities and professionals who advise women that in the interest of their child they should stick to the rule of not drinking one single glass in pregnancy at all (Health Council 2005; NHS 2013), justify this policy as a measure of precaution, in light of the fact that a safe limit has not been established. One glass may easily lead to more and thus into the danger zone. Moreover, it may well be that further scientific research will reveal that light drinking is less harmless than presently thought, or that harmful effects may occur at different levels of drinking for women with different types of genetic make-up (Lewis et al. 2012). But if these are good reasons to prefer erring on the side of caution, that is still something that women should be allowed to decide for themselves on the basis of accurate and balanced information. There are also consequentialist arguments for not suggesting more evidence of harm than there actually is: it may lead to unfounded guilt feelings in women who did not stick to the no-alcohol rule; it may (if found out) also further undermine the already shaky trust of the public in professional health advice (Gavaghan 2009).

8.3.5 Subsidiarity

The fifth and final condition is that of subsidiarity. This means that with regard to measures ranging from mild pressure to outright coercion always the least drastic measure must be chosen that would still effectively secure the end of protecting the future child from harm. Strong forms of pressure (involving the promise of bonus or the threat of a penalty) should only be considered if explicit directive counselling remains without the desired effect, and coercion only *ultima ratio*, meaning that there are no less invasive means to effectively protect the future child from harm. For instance, one may ask if the Rotterdam protocol involving routine requests for legal supervision (briefly described above as an example of strong pressure) can be justified in the light of this criterion (Boonekamp et al. 2012). Clearly, in the context of pregnancy, there may be time constraints that need to be taken into account. Exploring less invasive alternatives may require time that in cases of concern about prenatal exposure will often not be available, as it may entail missing the momentum for taking measures that could still be effective and proportional. If concerns are about perinatal, neonatal or postnatal harms, such time-constraints will be less of an immediate problem.

8.4 Competence and Paternalism

With regard to the ethical justification of pressure and coercion, much depends on whether the pregnant woman can be considered competent, in the sense of being able to make informed decisions about the issue at hand (Faden et al. 1986). The issue at hand is life-style adaptation during pregnancy. If an addicted pregnant woman is incompetent to make the relevant decisions, the justification of pressure or coercion may also be a matter of protecting her against herself. The underlying assumption is that it is not in her interest to give birth to a child that is severely damaged and/or will be taken away from her to be raised by foster parents. Given that she is unable to evaluate her continued drug abuse in the light of these possible consequences, paternalistic interference aimed at preventing these outcomes may be justified. In such cases, it can be said that the principle of beneficence (requiring professionals to serve the best interests of their patients or clients) prevails over respect for autonomy. However, this need not to be seen as a clash between opposite principles, as one may also argue that the aim is to help the woman in question to get her life on track again: paternalism is then conceived as a means to restore the capacity of meaningful self-determination (Verkerk 1999). If the woman is indeed incompetent, measures of pressure and coercion would still need to be justified in terms of the conditions discussed above. But the outcome may be different, as the weight of her own interests in having a healthy child may change the balance of proportionality. This means that in case of incompetency, more drastic measures including forced hospitalization may be more easily justified than in case of drug abusing women who are not incompetent. At this point, it is important to note that addiction as such does not equal incompetence (Berghmans 2000). Whether a person is incompetent will have to be determined on a case by case basis (Faden et al. 1986).

8.5 Legal Title for Coerced Hospitalization

Coerced hospitalization (to ensure abstinence or controlled intake, or to avoid untreated NAS) not only requires ethical, but also legal justification. In the Netherlands, the legal title for this is the Act governing 'special admissions to psychiatric hospitals'. This Act (BOPZ, Bijzondere Opnemingen in Psychiatrische

Ziekenhuizen) stipulates that a person can only be hospitalized against his will, if this person suffers from a psychiatric disorder that causes a 'danger either for the person him or herself or for another'. Dutch jurisprudence shows that in the context of this Act, 'another' may also be taken to refer to the unborn (but viable) child (Hondius et al. 2011). Whether in concrete cases the 'danger' is sufficiently great to justify coerced hospitalization is for the judge to decide. In a recent ruling, a license was given by the Court of Amsterdam to hospitalize a cocaine-addicted pregnant woman (at 16 weeks of pregnancy) because of the perceived danger that in case of continued abuse she would 'rob her unborn child of its life or seriously harm it'. The case is interesting for several reasons (see also below), including the absence of other psychiatric disorders than the woman's cocaine dependency, qualified in the psychiatric expert report as 'predominantly determining her cognition, affection and purposeful action' (Rechtbank Amsterdam 2011). This qualification is important as it distinguishes mere addiction from the kind of condition that would be needed for a successful appeal to the BOPZ: an addiction that seriously impacts upon the cognitive, emotional and behavioural faculties of the person in question ('addiction plus') (Höppener et al. 2013). As this qualification would not necessarily apply to women who 'only' continue drinking too much (one or more units of alcohol per day), a legal title for measures of coercion to protect their future children from a high risk of serious harm, may not be available in concrete cases of alcohol abuse.

8.6 Prenatal Child Protection and the Scope for Abortion

How do measures for prenatal child protection relate to women's right to have an abortion? In the Netherlands, where termination of pregnancy is legally allowed up to viability (specified as 24 weeks of pregnancy), it has been suggested that coercive measures to protect the child from harm cannot be taken prior to this limit, as that would undermine the scope for legal abortion by making the non-viable fetus an independent subject of legal protection (Bijlsma et al. 2008). And indeed, with the above court case as a remarkable exception, judges have not until now licensed coercive hospitalization prior to 24 weeks of pregnancy. Care professionals have protested that if meant to avoid harm resulting from exposure to prenatal drug abuse, this is simply too late (Schneider et al. 2004). However, this legal deadlock only arises if it is assumed that protecting the (as yet) unborn child equals protecting the fetus. Only on that assumption is it the case that taking protective measures prior to the abortion limit would make it difficult to at the same time maintain that women still have the right to choose abortion. However, the problem disappears if, as we have suggested, prenatal child protection is not about protecting the fetus, but about protecting the health and wellbeing of the future child. As such, the future child does not yet exist, but the interests that it will have after birth can already be harmed during pregnancy (Murray 1987; Den Hartogh 2010). Irrespective of one's views about the status of the fetus at different stages of development, measures of pressure or coercion to protect the future child can be justified at any stage of pregnancy without detracting from the pregnant woman's right to have an abortion. But if she intends to carry the pregnancy to term, this entails a responsibility towards the future child that professionals may remind her of and that in cases of neglect may constitute a ground for prenatal child protection measures, also prior to viability (Den Hartogh 2010).

Interestingly, the Amsterdam Court in the above case seems to have mixed up these different lines of reasoning (Rechtbank Amsterdam 2011). By licensing coercive hospitalization because of a danger for the unborn child already at 16 weeks, the court at first sight seems to acknowledge that if it comes to protecting the interests of the future child, 'the timing of harm is indifferent' (Murray 1987). However, if the court were indeed concerned about the future child rather than the fetus, it can be asked if the risks associated with cocaine abuse are substantial enough to justify coerced hospitalization. As argued above, this is highly questionable. Looking at the case report, it is clear that instead the court was mainly concerned that continued cocaine abuse would lead to fetal death. It was pointed out that the woman in question had already lost four pregnancies allegedly as a result of cocaine abuse (the report does not tell us what evidence was considered for suggesting a causal relation at this point), and that if she were not coerced into abstinence, there was a clear danger of the same happening again. Given that in Dutch Law, it is only at birth that a new human being becomes a person, it is at least remarkable that the court would go so far as license the coerced hospitalization of an addicted but not incompetent woman in order the protect the fetus from dying in utero. But it is even more remarkable that the court would do so to save the life of a fetus that has not yet reached the threshold of viability, until which under Dutch law abortion is legally possible. As the court seems to have been concerned about the life of a pre-viable fetus rather than about the wellbeing of the future child, its ruling seems difficult to reconcile with women's right to have an abortion.

8.7 Conclusion

Taking account of the vulnerability of those whose interests are at stake: addicted pregnant women (who are often also psychiatric patients) and their future children, it is of great importance to make the considerations behind the use of pressure or coercion explicit. Only thus will it be possible to avoid lapses in two directions: presenting women's right to self-determination as always trumping the interests of the future child, and suggesting that measures infringing upon women's freedom to lead their own lives need no further justification if the aim is to protect the future child. As a practice aid, we conclude this paper by presenting three types of questions that professionals in prenatal care for addicted pregnant women must ask themselves before deciding about taking measures of pressure or coercion in the interest of the future child:

1. What are the facts? Does the woman's lifestyle or behaviour entail a real risk or danger for the health or wellbeing of the future child? What (further) information is needed to answer this question?
2. Should measures of prenatal child protection be taken? Is the risk so large that professionals should try to protect the future child if the woman is unable or unwilling to assume her responsibility for the health and wellbeing of her future child?
3. How can this be done in an ethically responsible way? What is the least invasive measure on the escalation ladder of measures of pressure or coercion that would enable effective child protection? Is the relevant measure ethically acceptable in light of the level of interference in the woman's life?

Finally, it is important to note that coercion requires a legal title. The concern that this would only be possible in later stages of pregnancy in order not to jeopardize women's right to have an abortion, mistakenly assumes that prenatal child protection is about the fetus. However, it is about the future child.

Contributor Statement Wybo Dondorp and Guido de Wert have equally contributed to the contents of the paper. The paper was drafted by Wybo Dondorp.

References

Bakker, R., L.E. Pluimgraaff, E.A. Steegers, H. Raat, H. Tiemeier, A. Hofman, and V.W. Jaddoe. 2010. Associations of light and moderate maternal alcohol consumption with fetal growth characteristics in different periods of pregnancy: The Generation R Study. *International Journal of Epidemiology* 39: 777–789.

Berghmans, R.L.P. 2000. *Bekwaam genoeg? Wils(on)bekwaamheid in geneeskunde, gezondheidsrecht en gezondheidsethiek.* Utrecht: Nederlandse Vereniging voor Bio-ethiek.

Bijlsma, M.W., J.M.B. Wennink, A.C. Enkelaar, M.H.B. Heres, and A. Honig. 2008. De mogelijkheid van ondertoezichtstelling van het nog ongeboren kind bij twijfels over de veiligheid van de thuissituatie. *Nederlands Tijdschrift voor Geneeskunde* 152: 895–898.

Boonekamp, J., R. Berghmans, W. Dondorp, and G. de Wert. 2012. Zorg voor verslaafde zwangere vrouwen: rechtvaardiging van drang en dwang. *Tijdschrift voor Psychiatrie* 54: 257–266.

Buckingham-Howes, S., S.S. Berger, L.A. Scaletti, and M. Black. 2013. Systematic review of prenatal cocaine exposure and adolescent development. *Pediatrics* 131: e1917–e1936.

Cressman, A.M., A. Natekar, E. Kim, G. Koren, and P. Bozzo. 2014. Cocaine abuse during pregnancy. *Journal of Obstetrics and Gynaecology Canada* 36: 628–631.

Den Hartogh, G. 2010. Prenatale en postmortale schade: Temporele grenzen van rechtssubjectiviteit. *Nederlands Juristenblad* 645: 778–783.

Dondorp, W., and G. de Wert. 2015. Drang en dwang: ethiek van prenatale kinderbescherming. In *Handboek psychiatrie en zwangerschap*, ed. M. Lambregtse-van den Berg, I. Van Kamp, and J. Wennink, 463–469. Utrecht: de Tijdstroom.

Faden, R.R., T.L. Beauchamp, and N.M. King. 1986. *A history and theory of informed consent.* New York: Oxford University Press.

Gavaghan, C. 2009. 'You can't handle the truth': Medical paternalism and prenatal alcohol use. *Journal of Medical Ethics* 35: 300–303.

Health Council of the Netherlands. 2005. *Health Council of the Netherlands. Risks of alcohol consumption related to conception, pregnancy and breastfeeding*. The Hague: Health Council of the Netherlands.

Hondius, A.J.K., T.E. Stikker, J.M.B. Wennink, and A. Honig. 2011. Wet BOPZ toegepast bij vroege zwangerschap van verslaafde. *Nederlands Tijdschrift voor Geneeskunde* 155: A3818.

Höppener, P.E., J.A. Godschalx-Dekker, and B.J.M. Van de Wetering. 2013. Gedwongen opname bij stoornissen in of door het gebruik van middelen. *Tijdschrift voor Psychiatrie* 55: 269–277.

Kelly, Y., M. Iacovou, M.A. Quigley, R. Gray, D. Wolke, J. Kelly, and A. Sacker. 2013. Light drinking versus abstinence in pregnancy, behavioural and cognitive outcomes in 7-year-old children: a longitudinal cohort study. *BJOG: An International Journal of Obstetrics and Gynaecology* 120: 1340–1347.

Kocherlakota, P. 2014. Neonatal abstinence syndrome. *Pediatrics* 134: e547–e561. doi:10.1542/peds.2013-3524.

Lewis, S.J., L. Zuccolo, G. Davey Smith, J. Macleod, S. Rodriguez, E.S. Draper, M. Barrow, R. Alati, K. Sayal, S. Ring, J. Golding, and R. Gray. 2012. Fetal alcohol exposure and IQ at age 8: Evidence from a population-based birth-cohort study. *PLoS One* 7: e49407.

Murray, T.H. 1987. Moral obligations to the not-yet born: The fetus as patient. *Clinics in Perinatology* 14: 329–342.

National Health Service. 2013. Alcohol in pregnancy. What's the official advice? www.nhs.uk/conditions/pregnancy-and-baby/pages/alcohol-medicines-drugs-pregnant.aspx#close. Last accessed June 2015.

Okie, S. 2009. Crack babies: The epidemic that wasn't. *The New York Times*, 26 Jan 2009.

Rechtbank Amsterdam. 2011, 14 Februari. Zaaknummer 482593/11.1200 F http://uitspraken.rechtspraak.nl/inziendocument?id=ECLI:NL:RBAMS:2011:BW7232. Last accessed June 2015.

Robinson, M., W.H. Oddy, N.J. McLean, P. Jacoby, C.E. Pennell, N.H. de Klerk, S.R. Zubrick, F.J. Stanley, and J.P. Newnham. 2010. Low-moderate prenatal alcohol exposure and risk to child behavioural development: A prospective cohort study. *BJOG* 117: 1139–1150.

Schaefer, C., P.W.J. Peters, and R.K. Miller. 2007. *Drugs during pregnancy and lactation*. Amsterdam: Academic Press, Elsevier.

Schermer, M.H.N. 2003. Drang en informele dwang in de zorg. In *Raad voor de Volksgezondheid & Zorg, Signalering Ethiek en Gezondheid*. Den Haag: Centrum voor Ethiek en Gezondheid.

Schneider, A.J., M. Raats, M.J. Blondeau, and E.A. Steegers. 2004. Zwangere, verslaafde prostituees: soms gedwongen opname in het belang van het kind. *Nederlands Tijdschrift voor Geneeskunde* 148: 1949–1952.

Verkerk, M.A. 1999. Care perspective on coercion and autonomy. *Bioethics* 13: 358–368.

Chapter 9
The Confused Stork: Gender Identity Development, Parental and Social Responsibilities

Simona Giordano

Abstract From the moment a child is born, what happens to that child will be heavily based on gender predictions. What usually influences gender predictions is genital morphology. However, endocrinology and developmental psychology show that genital morphology is not always an accurate predictor of gender identity. This chapter reviews some of the recent literature on sex and gender identity development and shows that there seem to be not only two, but many sexes and genders. One related fact is that there is no set of biological markers that can allow us to determine whether an individual is a female or a male. If we accept these facts, we are led to ask how society as a whole should be organised, how parents should raise their children, how nurseries and schools should operate, and how society at large should function. This chapter does not advocate the abolition of sexes and genders: it instead proposes that they should be treated as broad approximations and illustrates some ways in which tolerance towards sex and gender minorities may be fostered.

Keywords Gender identity • Sex differentiation • Transgender minors • Gender • Sex • Intersex

> [To the new generations]…I should like to say two things, one intellectual and one moral.
>
> The intellectual thing I should want to say to them is this: When you are studying any matter, or considering any philosophy, ask yourself *only* what are the facts and what is the truth that the facts bear out. Never let yourself be diverted either by what you wish to believe, or by what you think would have beneficent social effects if it were believed. But look only, and solely, at what are the *facts*. That is the intellectual thing that I should wish to say.

S. Giordano (✉)
School of Law, The University of Manchester,
Williamson Building-2.27, M13 9PL Manchester, UK
e-mail: simona.giordano@manchester.ac.uk

K. Hens et al. (eds.), *Parental Responsibility in the Context of Neuroscience and Genetics*, International Library of Ethics, Law, and the New Medicine 69,
DOI 10.1007/978-3-319-42834-5_9

> The moral thing I should wish to say to them is very simple: I should say, love is wise, hatred is foolish. In this world which is getting more and more closely interconnected, we have to learn to tolerate each other [...] We can only live together in that way — and if we are to live together and not die together, we must learn a kind of charity and a kind of tolerance, which is absolutely vital to the continuation of human life on this planet. (Bertrand Russell 1959)[1]

9.1 Introduction[2]

Once upon a time, the Stork, flapping her wings, used to carry babies in blue or pink cloths, leaving blue or pink ribbons for parents to pin up their front doors: expecting families would rejoice in receiving their baby boys and girls. Babies who were not clearly boys or girls (for example, those whose genitals were ambiguous) were probably delivered in surgical units, where the anomalies could be corrected, so that they could be assigned to a sex, registered, redelivered to their homes and, finally, treated and reared as boys or girls.

From the moment a child is born, what happens to that child will be heavily based on gender predictions. The child's sex morphology is likely to influence those predictions. In other words, if a baby has a penis usually it will be considered a boy, as such *he* will be registered and as such raised. It will normally be expected that at some point the boy will describe himself as a boy, will be happy with the boy-name given to him, with the boys-clothes that he'll be expected to wear, and that this sense of congruence between 'having a penis' and 'feeling a boy' will remain unaltered along his life: one day, the boy will be a man. A similar set of gender predictions is likely to apply to babies born with a vagina.

However, endocrinology and developmental psychology tell us at least two important facts: the first is that there are no clear sets of biological markers that can be used to determine whether individuals are boys or girls. As we shall see later in this chapter, it is not clear what it is exactly that makes a zygote develop into a person with a penis and testis and another into a person with a vagina and ovaries. The second fact that endocrinology and developmental psychology tell us is that there are not only two sexes in humans: there are many gradations, many sexes on the male-to-female spectrum. Unsurprisingly, there are many more than two genders too. Many people *feel* like a man or a woman (whatever that means to them), many people feel a bit of both, many others feel neither, while many feel like a man in some circumstances and a woman in others.

It also appears that genital morphology, genetics and chromosomal make-up are not reliable predictors of gender identity. One person could have a vagina and an X chromosome, and yet *know (or feel, having the certainty)* that he is a boy – just born 'in the wrong body' - and *viceversa*. Why this happens is not known: but it is equally

[1] www.brainpickings.org/index.php/2012/05/24/love-is-wise-hatred-is-foolish-bertrand-russell-1959/ Accessed February 2016.

[2] I wish to thank John Harris for reading and commenting on this chapter.

unknown why the same person may *know (or feel, having the certainty) that she is a girl* – same for the boys. The way in which gender identity develops is likely to be extremely complex and it is not fully understood.

This leaves our Stork very confused. How should she decide what ribbon to use? What should parents be told? Should the colour codes be dropped, and should the Stork just deliver babies 'as babies'? And what should parents (all parents of course) do, if they cannot be sure whether they are raising a boy or a girl? Would it be best if parents chose gender neutral names? Or gender neutral clothes? How should nurseries and schools operate? How should society as a whole be organised? And ultimately, why does all this matter (if it does)? In order to address these questions, I will first provide a brief account of some theories of gender identity development. In the course of this account, it will become clear that the 'facts' around gender identity development are extremely complex, and understanding these complexities will help us to answer at least some of these questions. Thus let this analysis begin by looking at how gender identity development, according to some research, is thought to occur.

9.2 Beth and Adam: The Role of Parents in Shaping Gender

Usually 'sex' and 'gender' are differentiated in the literature as follows: sex is regarded as the set of biological facts (chromosomes, hormones, gonads, genitalia) and gender as the social interpretation of these facts (what it means to have a certain physical endowment in a given society at a certain point in time). I will return in more detail to these notions later and discuss the problems involved in this differentiation. For now I am going to give an account of the studies that have attempted to understand at what point and how children acquire a sense of being 'a girl' or 'a boy', and how this happens. That sense of identification (which usually is expressed in the form of 'I am a boy' or 'I am a girl') is usually called 'gender identity'. In the following, I will use these terms in line with these typical uses.

One theory of gender identity development is the one that suggests that gender identity is acquired, and is mainly 'a social construct': from this point of view it is parents, or significant others, that have a primary, major role in the 'construction' of a child's gender identity. Various studies, as we are now going to see, corroborate this theory. Firstly, some studies show that, when relating to babies and young children, adults behave differently depending on the sex they believe the baby or child to have: in one study, for example, a group of young women were observed while interacting with a baby, Beth, aged 5 months. They were seen smiling often to the child and offering her dolls to play. The girl was said to be 'sweet'. Then a group of different young women were observed while interacting with a boy, called Adam, also aged 5 months. The women in this case were offering Adam toy trains and showed reactions that were remarkably different to those of the previous group (Will et al. 1976). Beth and Adam were the same baby, only dressed differently (Giddens and Griffiths 2006: 170). As Green summarises, these and other experiments illustrate that people act differently around children, if they believe that these children are boys or girls (Green 2009). According to Archer and Lloyds, from the time that parents learn whether the new baby is a boy or girl, many aspects of the way it is treated will be influenced by its sex (Archer and Lloyd 2002: 60–71).

Secondly, it has been shown that children are motivated to talk and act in ways that are seen within their social group as appropriate to the birth sex. In other words, whether or not they realise it, parents and significant others model gender behaviour, encourage children to behave 'appropriately', and reinforce them when they do (Lorber 1994). The child responds with feelings and behaviours that are congruent to these triggers (Ruspini 2009: 73). Bandura noted that parents in many cultures provide play experiences that are sex typed, and children, he argued, *learn to behave differently because of these experiences*. Bandura coined the term *social learning theory*. Children monitor their behaviour against the expected standards and feel pride in performing gender role consistent behaviour, even if there is no explicit external sanction or praise (Bandura 1965). Later studies seem to confirm these earlier findings (Devor 1989).

Finally, once the gender is accepted by the child, and once gender categories have been internalised, children also apparently begin *to disapprove* of those who behave in a gender non-congruent manner. In one study, Bussey and Bandura asked nursery children aged three and four to evaluate gender typed behaviour of peers from videotapes. They showed girls playing with 'masculine' toys and boys playing with 'feminine' toys. The children regularly showed disapproval of gender inconsistent behaviour (for example, boys playing with dolls) (Bussey and Bandura 1992). Lloyd and Duveen studied 120 children aged 18 months to 3 years old and arrived at similar conclusions (Lloyd Duveen 1990).

Put in simple words, under this perspective children are not *born* as boys and girls: they may be born with male or female genitalia, or with male or female chromosomes, but *they become boys and girls*, depending on how they are treated, and because of the way in which they are treated. On this view gender is not simply or even strictly a biological fact: gender is a social construct, and the primary actors in this construction are the parents and/or the people close to children in the early months and years. Before I provide an account of a very different theory of gender development, it is important to reflect on some of the implications of this theory.

9.2.1 Implications of the Theory

The belief that children are *not born* as boys or girls, but *become such*, depending on how they are treated, particularly early on by the significant others, has widespread implications. One is that babies born with intersex conditions (where, for example, the genitalia are ambiguous) have been routinely subjected to surgical operations as infants. A vagina would typically be created, regardless of the chromosomal make-up. Under this perspective, since gender is *learned*, if a child is raised as a girl, that child will eventually become a girl. The practice of routinely subjecting infants to medically unnecessary genital surgery has been widely debated and criticised in the literature, but is still common (Vidal et al. 2010).

But there are other implications as well. The idea that gender is constructed by society, rather than 'a fact of nature', had particular importance in feminist debates, in particular in movements for the liberation of women. Claims such as that *by nature women are more suited for childrearing*, for example, or that *by nature girls are more nurturant* than boys, or that *men are better suited for certain types of jobs*, have a precise political meaning. Similar claims, as already John Stuart Mill noted in 1869, have been used historically to 'justify' the subjugation of women, as well as the slavery of black people in America (Mill 1869): the black are *by nature* better suited for physical work, the woman is *by nature* better suited for house work. Scott, on this line, argued that 'genders' *deny people the chance to be who they really are* (Scott 1986). If this is right, until we wonder whether a baby is a boy or a girl, or whether a person is a man or a woman, we somehow are on the wrong page.

From what has been said so far, one could be led to believe that referring to babies as either boys or girls may deprive them of an open future; may encapsulate them in fixed roles that may not reflect their identity; may limit their capacity to explore who they are and what their more genuine inclinations are. Genders are straightjackets to one's identity, to one's flourishing, under this perspective. Although, at first sight, these studies are compelling, other studies on gender identity development present a rather different picture. Let us then move to a rather different account.

9.3 The Theory According to Which Gender Identity Is Innate

Bowlby was one of the first psychologists to have argued that gender identity is hardwired in individuals, and is thus primarily the result of biological factors (Bowlby 1969). On this line, Simon Baron-Cohen wrote: "The female brain is predominantly 'hard-wired' for empathy. The male brain is predominantly 'hard-wired' for understanding and building systems" (Baron-Cohen 2003: 1).

Bowlby observed a number of children, and saw repeatedly that their preferences for toys and activities depended on their gender. He also noted physical differences: male babies are generally bigger; boys often sleep less and cry more, and are generally more active, whereas girls start talking earlier than boys, and so on (Bowlby 1969). The literature on sex differences in infants and young children is abundant. In one study, babies aged 18 months were presented with some pictures of faces of infants of the same sex and of the other sex. It was found that boys looked at boys' faces longer and girls looked at girls' faces longer. This was regarded as an indication that, even at this early stage, babies have some recognition of "like me" and "not like me" (Green 2009). It also appeared in this and other studies that boys and girls begin to prefer 'sex-typed' toys by the age of 1 year, and that around the age of three and a half and four and a half boys prefer to play with boys and girls with girls

(Green 2009; van De Beek et al. 2009; Lamminmaki et al. 2012). It could be asked whether some of these gendered behaviours may result from social clues that operate very early in the life of the child, even before the child is born, as discussed earlier. Some studies however show that at least some of these differences are probably mainly biological and innate.

Nelson reports that research on animals (for example, on birdsongs, on urinary posture in canines, even in fish, and in mammalians such as rhesus monkeys) suggests that sexually dimorphic behaviour has a strong relationship with brain structure and hormones, including prenatal exposure to hormones (Nelson 2005, chapter 4, in particular 230–232). Nurturing ('maternalism'), affiliation (nonsexual peer relations), aggression and activity levels, all of which show normative sex differences (including in non-human animals) appear to be affected by experimental manipulations in exposure to prenatal sex hormones, including androgens (Zucker 2001: 110). If these results can be extrapolated and applied to human behaviours, they seem to suggest that at least some gender roles (maternalism or aggression for example) do result at least to some extent from prenatal hormonal exposure, and thus are not primarily the result of parental behaviour.

Recent studies on humans seem to lend some support to this hypothesis. These studies show that the neurological development of 'boys and girls' differs significantly. So, 'sex differences' are, according to these studies, present in humans: for example the total brain volumes and grey matter volumes are different in boys and girls (Burke 2014: 16, Ingalhalikar et al. 2014), and this difference, present at prepubertal level, increases with pubertal development (Goddings et al. 2013). This means that hormonal exposure 'modifies' the brain structure, and this is highly likely to in turn modify cognition, emotion and therefore behaviour. Visuo-spatial cognitive functioning also seems to differ in males and females (both children and adults), and such difference seems to be due to the effects of testosterone (Burke 2014: 125–43). Again, hormone exposure modifies the way in which we 'see' the world and elaborate it. From this perspective, gender behaviour and gender identity are likely to develop, at least to some important (yet not quantifiable) extent, independently of how parents or significant others act with their babies and children.

A recent study on click-evoked otoacoustic emissions (CEOAEs) provides further interesting data on the relationship between biological differences among sexes and gender identity. CEOAEs are echo-like sounds produced by the inner ear in response to click-stimuli (Burke 2014: 25). Women and men have different CEOAEs emissions. There is thus a clear, observable sex difference in the way men and women respond to acoustic stimuli. Women usually have stronger responses compared to men, and this difference is also found in neonates. Baby girls thus have stronger CEOAEs responses than baby boys.

Because baby boys and baby girls *already* manifest this sex difference, it is likely that the weaker responses found in neonatal boys depend on higher exposure to testosterone during prenatal life compared to baby girls. Burke et al. have examined the CEOAEs responses in 24 natal boys and 23 natal girls with gender dysphoria (boys and girls who perceive their gender *as incongruent* with their body) and 65 boys and 62 girls with no gender issues. Thus, all in all, they have tested several hundred ears. They found that *boys with gender dysphoria had more female-typical CEOAEs emissions*. In other words, their hearing response was more similar to 'girls' than to 'boys'. Girls with gender dysphoria, instead, had emissions similar to girls without gender issues. This seems to suggest that boys with gender dysphoria may have been exposed to lower amounts of androgen during early development, compared to boys with no gender issues (thus, one suggestion here is that gender dysphoria, being transgender at least in natal boys, may be due – perhaps among other things – also to a lower than usual exposure to androgens). But this also seems to suggest something else, which is very important for the purposes of this chapter: if CEOAEs emissions are linked to prenatal hormone exposure, this means at least the following:

1. That *some sex differences* are not socially constructed, do not necessarily depend on the behaviour of parents or significant others, and begin before birth (how we should think about these sex differences, and how families, society at large, and even medicine should deal with them is another matter);
2. That *at least in some cases gender identification is congruent with sex differences that are not immediately evident* – simply put, the gender of a transgender girl (a natal boy who feels she is a girl) is congruent with her acoustic emissions, and incongruent with her more evident sex difference - genitals (penis). Thus perhaps *some sex differences* are more telling about gender identity than those usually taken as being indicative of it (the X Y chromosomes, or the genitals);
3. That it is likely that *gender identity* (and *not only sex differences*) relates to some biological facts (such as androgen exposure), some of which may occur prior to birth.

The complicating factor is that no difference in CEOAEs emissions was found in girls with gender dysphoria compared to girls with no gender issues. Thus if lower androgen exposure may cause boys to have CEOAEs emissions more similar to girls, and may even be a contributing factor to the development of a female gender identity, the reverse may not be true: it may not be true that the acquisition of a *male gender identity* (not of a male physique of course) has anything to do with greater exposure to androgens prenatally (otherwise, natal girls with gender dysphoria would have CEOAEs emissions more similar to boys, weaker than the norm for girls).

This perhaps unexpected finding is coherent with previous studies on the length of the fingers. Earlier studies examined the ratio between the length of the index and the ring finger, known as 2D:4D ratio. Men generally have lower 2D:4D ratio compared to women (Williams et al. 2000; Grimbos et al. 2010; Honekopp and Watson 2010). This is present in children and also prenatally (McIntyre et al. 2006). Boys with gender dysphoria have 2D:4D ratio *more similar to females than to males* (similarly they have CEOAEs emissions more similar to females than to males). But girls with gender dysphoria have 2D:4D ratio similar to all other females. Thus again, lower androgen exposure may cause *the sex difference*, and may also contribute to the development of a *female gender identification in natal boys*; but because women with gender dysphoria have 2D:4D ratio similar to women without gender issues, it cannot be hypothesised that the masculinisation of gender identity in girls is due to higher than normal androgen exposure.

There is a further complication: some sex differences seem to be related to sexual orientation, rather than gender identification. Otoacustic emissions (OAEs), for example, *in lesbians* appear similar to those of *heterosexual males*, regardless of their gender identification. But *gay men* also have emissions *similar to heterosexual males*. The relationship between gender identification, sexual orientation and hormonal exposure is thus not yet clear.

The response to odorous steroid has also been studied; the hypothalamic response to odour in boys and girls differs, and people (in this case adolescents were the subjects of the study) with gender dysphoria were found to respond to odour in a way that is more similar *to their experienced gender*. In other words, adolescents who are born as boys but who perceive themselves as girls have hypothalamic response to odours that is more similar to natal girls; the same applies to girls with gender dysphoria (Burke et al. 2014).

This again suggests that some differences between sexes are not induced by parents or by society at large: there seem to be biological differences between sexes that are not constructed by the family or the society in which the baby is raised. Moreover, this suggests that some not obvious differences (such as the hypothalamic response to odour or the otoacustic emissions) may be more closely related to a person's gender than their chromosomes or genitalia. Genitalia and internal organs such as reproductive organs or chromosomes may thus not be the most accurate markers of one person's gender. This could also indicate that there may well be women born with a penis and testis and men born with ovaries and vagina; these people are now regarded as *transgender* because it is believed that their gender is not congruent with their sex features. But it may well be that women with a penis and men with a vagina may not be transgender at all, and may instead have a gender that is indeed congruent with *some* of their sex features (just not the genitalia).

9.3.1 Further Considerations

The research reported above could be interpreted in at least two ways. It could indicate that humankind is constituted of males and females; that if you wish to make (or adopt) babies you may expect to have a boy or a girl, and that these two groups are characterised by some identifiable features. These features result partly from the chromosomal makeup, and partly from hormone exposure. Gender identity and sexual orientation intersect with physical development (sex differentiation) in a difficult manner, one that still needs to be understood, but at least it is clear that babies and people are either of a female or of a male sex, and those who are neither are affected by some (again identifiable) disorder or medical condition.

The other interpretation is that there are different ways in which humans may develop, not only in terms of their gender identity, but even in terms of their biological makeup; traits that are not found in some individuals are found in others (for example male typical CEOAs emissions may be found in individuals with an XX set of chromosomes and female phenotype). Some sex differences are found in foetuses and it is thus unlikely that they are induced by external clues; some of these differences are associated with atypical or non-conforming gender, which also suggests that gender identity development (whether or not conforming) somehow relates, in some non-fully understood way, to prenatal development. Thus, rather than suggesting that there are two sexes and two genders, and then an array of anomalies in the middle, this research can also be interpreted as showing that there are indeed several sexes and several genders, that humankind is composed of not just males and females, but of a very diversified array of combinations.

In the next section we shall discuss sex differentiation further and offer support to this second interpretation.

9.4 Sex and Gender

As anticipated in the introduction, a significant part of the literature on gender (including the literature which criticises biological determinism) differentiates between *sex and gender*. Sex is thought to refer to a number of biological 'facts'. First there is the 'chromosomal sex' (XX; XY); then at about 6 weeks gestation, we have the expression of so called 'gonadal sex'. At this stage, the tissues that are now developing in the embryo may either produce androgen hormones (usually where an XY is present), or may not (usually where an XX is present). Accordingly, typically male or female internal and external organs will be formed. This is the stage of 'hormonal sex', which induces differentiation at the second month of foetal life. Lastly, there is the 'anatomical sex', which refers both to the internal reproductive organs and external phenotypical appearance (genitalia) (Gross 2010). As we have seen earlier in this chapter, other facts are usually taken as 'sex differences': size, sleeping patterns, and so on.

Gender is usually seen as the "social interpretation" (Gross 2010: 563) of these biological facts. Gender identity refers to the acquisition of roles, behaviours, preferences that match, or are supposed to match, expectations that in various societies are attached to being a biological boy or a girl. For an individual, gender identity is the sense of belonging to one group (for example, to the boys), with all that this encompasses.

What it is to be a man or a woman clearly changes throughout history; it changes across the world and it depends on cultural, religious, moral norms, economic and political settings and perhaps many other variables. Thus gender is often regarded as 'the difficult one'; the one that has moral and political implications, the one muddied by social norms. Sex, according to this perspective, is instead the set of the observable, objective *data*. Sex can just be observed, it is not a matter of interpretation, from this point of view.

The problem is that sex differentiation is not as simple as illustrated here (Vernon 2009: 269).[3] We have already seen in the sections above that the sex differences found in humans and other mammals are many and varied, and their interaction with gender identity development is complex and scarcely understood. But let us look more closely at sex differentiation.

Originally it was thought that the baby's sex depended on the presence of the X or Y chromosome at the 23rd pair (XX female, XY, male). In the 1990s a gene determining the differentiation of tissue into testis was identified within the Y chromosome. This gene was called SRY (sex determining region of the Y), also known as TDF (testis development factor). It was thought that the SRY triggers a pathway of other genes that cause the gonads to continue to develop into a male. However, it was also found that there are SRY negative individuals who have testicles and SRY positive individuals who do not have testicles. In these latter cases, there seems to be another gene (this time within the X chromosome) called DAX-1, which can override the effects of the SRY, "so that an individual with XY chromosomes and a functioning SRY gene develops ovaries and not testes" (Newbould 2015). There seem to be at least 12 other chromosomes across the human genome governing sex differentiation.[4]

Anne Fausto-Sterling thus argues that "biologically speaking, there are many gradations running from female to male; and depending on how one calls the shots, one can argue that along that spectrum lie at least five sexes—and perhaps even more" (Fausto-Sterling 1993; Feinberg 1996: 101).

The Intersex Society of North America reports:

> [N]ature presents us with sex anatomy spectrums. Breasts, penises, clitorises, scrotums, labia, gonads—all of these vary in size and shape and morphology. So-called 'sex' chromosomes can vary quite a bit, too. But in human cultures, sex categories get simplified into male, female, and sometimes intersex, in order to simplify social interactions, express what we know and feel, and maintain order. So nature doesn't decide where the category of 'male' ends and the category of 'intersex' begins, or where the category of 'intersex' ends and the category of 'female' begins. Humans decide (Intersex Society of North America).

To contend that only two sexes are 'healthy' or 'biologically appropriate' seems to reveal the assumption of normative categories through which the biological 'facts' are read and interpreted. There is no empirical fact based on which we can differentiate between 'healthy' and 'pathological' sexes. Even in those cases in which medical intervention may be necessary to prevent medical consequences associated with some forms of intersex, it is not the ambiguous or atypical sex that is pathological, but some features that are associated with it.[5] The differentiation of sexes in healthy and ill, if it takes place, does so on the basis of a value judgment,

[3] I owe this observation to Melanie Newbould.

[4] I wish to thank Melanie Newbould for the references and the considerations briefly summarised in these lines.

[5] In some forms of intersex, tissue is highly susceptible to develop neoplasia, and it may be clinically indicated to remove it to avoid the risk of formation of tumours. There are other medical complications related to various forms of intersex. I owe this observation to Melanie Newbould.

Table 9.1 Relationship between gender and sex

Human societies	→	Gender as a form of social differentiation	→	The interpretation of observable facts in terms of sex differentiation
	Produce		Leads to	

not on the basis of empirical observation. Empirical observation only tells us that humans develop in many different shapes and forms. Empirical observation does not tell us that some of these shapes and forms are healthy and some are pathological.

Thus arguably the classification of sexes into males, females, and disorders, depends on an implicit norm of functioning constructed on gender norms. The biological classification of sexes can perhaps be best understood by reversing the relationship between sex and gender: rather than seeing gender as a construction based on biological sex, we should perhaps think of sex as a construction based on implicit norms relating to gender (see Table 9.1). (Lewins 1995: 35–37).[6] Sex is not a neutral set of biological attributes. The way we look at the biological facts is filtered by our assumptions relating to how people should be. The very biological facts that should provide the 'substrate' for social constructs are themselves a matter of interpretation and construct, which cannot be separated from the social and cultural values of the observers.

At this point our Stork has totally lost the plot. She cannot determine whether babies are males or females just by looking at their genitalia - as even the Stork has by now understood that a baby could have a vagina and have a Y chromosome; or that a baby could have a penis, and a Y chromosome but early in life state that she is a girl; and thus the Stork has probably recognised that there is no easy way in which she can label babies as boys and girls, neither if she looks at the baby's genitals, nor if she takes further endocrinological tests, nor if she looks at their finger ratios, nor if she looks at the brain scans, or at the hypothalamic responses to odours, nor if she looks at the otoacoustic emissions. All of these factors tell us *something* about the baby, but it is not entirely clear *what.*

But what is clear is that there are several possible combinations of sexes and genders, and that these are not always related to genital anatomy. Despite this, the idea that gender *must* relate to genital anatomy continues to determine the way in which children are raised and treated, even in cases in which it is clear that the gender identity of a child is not congruent with the genital sex. Important decisions relating to children are still made on the basis of such misunderstanding of what being a boy or a girl is. A US court case illustrates this point poignantly.

[6] Table 9.1 is readapted from Lewins (1995: 36).

9.5 Gender Identification in the Courts: When Parents Fight Over a Child's Gender

That parents often disagree about what is good for a child does not surprise. But that parents may have divergent views on whether a child is a boy or a girl, and how he or she should be treated, is less usual.

Such a disagreement was at the core of a US court case in 2007 (Smith v. Smith 2007) and concerns a child registered as a boy at birth, but who since the first years of life identified herself as a girl.[7] The parents divorced when the child was 6 years old, and the mother, Victoria, was awarded custody. Victoria was supportive of the child's perceived gender: the child was called Christine (a name that the child chose) and was allowed to wear girl's clothes (Smith v. Smith 2007, at 1). Mother and child also participated in transgender support groups. At the age of 9, Victoria attempted to enrol Christine in a new school as a girl. The father opposed the decision and brought the case to court. Victoria was ordered to stop calling the child Christine, and to rear the child as a boy. She was also ordered to stop any treatment or counselling for the gender dysphoria of the child, and to re-enrol the child in the original boys' school (Smith v. Smith 2007, at 7). Because she did not abide to the court's order, the father, Kevin, was thus awarded sole custody rights. At one point Christine emailed her father, enclosing photos of herself dressed in female clothing, and stating that 'God made a mistake' about her gender. She sent him a videotape in which she repeatedly stated that she was a girl (even if lacking all of a girl's 'body parts'), looked forward to wearing girl's clothes all the time, wanted to go to school as a girl, and to live a normal life as a girl (Smith vs. Smith 2007, at 13). Nonetheless the appellate court affirmed the initial ruling.

The decision was heavily criticised in the literature; it was questioned whether the experts involved had sufficient expertise in child gender identity development, and it was evident that the child's perception was not given much weight (Skougard 2012; Valentine 2009). One problem perhaps with giving priority to the child's view is that the majority of children who claim to be 'in the wrong body' as prepubertal children will *not* become transgender adults (Steensma et al. 2011). Most will become homosexual adults or lose their gender dysphoria as they mature and go on to live as gender typical heterosexual adults. So one worry may be that, by giving voice to the child, the child may begin the path of social transition maybe leading to later physical transition (involving extensive surgery and lifelong medical treatment), when the dysphoria would have subsided spontaneously in a 'less' encouraging environment.

Whereas the worry is understandable, there is no evidence that forcing the gender non-conforming child to conform to the natal sex is going to help him or her *to not become transgender*. Due to methodological and ethical issues, it is impossible to collect data of these sorts. However, the literature on transgender children consis-

[7] I wish to thank Dr Barry Lyons for referring me to this case.

tently shows that gender non-conforming children who are not allowed to express their perceived gender suffer great psychological harm. Conversion therapies are now regarded as useless at best, and harmful at worse.[8]

These facts matter not only to gender non-conforming children: they matter to all children and indeed to all of us, as they tell us something important about human nature and about how society at large should be organised in order to ensure the wellbeing and flourishing of each individual, particularly children, given that gender identity develops in the first few years of life (sex typing is thought to be complete by the age of five or six, Gross 2010: 571–573).

9.6 Why Does All This Matter?

There are various reasons why gender matters, and thus why it is important that people, and particularly parents, understand how it may develop. One reason is that how people are classified (boys or girls, men or women, or transgender or intersex) is likely to determine how they are treated not only within their families, but also in nurseries, in schools, in society at large (Gilligan 1993). Later on in their lives, marriage, employment, and many legal rights are often a function of 'gender' (Kitzinger 2001).

Gender and sex are also important *to people*, no matter how complicated their determination is. Whether or not their gender identity is conforming to birth sex, most people place themselves somewhere along the spectrum that runs from the concept of a boy to the concept of a girl. Not only that: most people need and some demand to be treated according to how they feel about themselves. *Gender* is so important to some people that they are willing to go through lifelong medical treatment and extensive invasive surgery to obtain a body that more closely resembles their experienced gender.

Misguiding ideas of what sex and gender are, and how they may develop, can hinder the health and flourishing of individuals. The idea that humankind is composed of boys, girls and people with disorders, has also obscured great violence suffered by sex and gender minorities. Routine genital mutilation of infants born with intersex conditions is still lawful in many countries, including England, and is performed solely on the basis of sex/gender binary ideas and related social norms of acceptability and normality. The blind acceptance of the categories of 'male' and 'female' thus risks overshadowing the routine performance of clinically unnecessary and ethically questionable genital surgery of infants, and subsequent upbringing which could be incongruent with the child's experienced gender.

The way in which gender binary stereotypes can be dangerous is also obviously illustrated by the case of lesbian-gay-bisexual-transgender (LGBT) people. People,

---

[8] See e.g. "Conversion therapy. Consensus statement", available at www.psychotherapy.org.uk/UKCP_Documents/policy/Conversion%20therapy.pdf. Accessed February 2016.

including children, who do not fit the gender divide are regularly subjected to high degrees of social ostracism, as well as to open physical violence or even murder (Wallien et al. 2010; Warwick et al. 2004). Studies show increased likelihood of homelessness in LGBT youths, school dropouts, high rates of verbal and physical abuse (including at school), increased likelihood of substance abuse and hepatitis B and C, as well as mental health concerns (Meininger and Remafedi 2008).

Obviously parents may be unable to protect children from these forms of discrimination and abuse, but being aware of the child's experienced gender, and being indeed open to the idea that children's sex and gender may develop in a variety of ways, may help to privilege more accepting environments for their children, and warn children of the possibility of peer discrimination and pressure. Being open to various sexes and genders may empower the children, by enabling them to express who they are, and thus to find support and protection. In societies in which the gender divide is not as marked as in ours, gender non-conforming people suffer less (Connolly 2003). An empirical study of the long-term psychological outcome of gender non-conforming children in different (more or less accepting) family environments has not, to the best of my knowledge, be performed to date, but it is not unreasonable to hypothesise that also in the more accepting families children are likely to suffer less. Correcting misguided ideas about sex and gender is obviously important for the protection of the sex and gender minorities. But we all need to bear in mind that the rigid sex/gender divide can harm each of us in many ways, whether or not we belong to or identify ourselves as a part of a gender minority. It can blind us to the complexities of the human condition, and it can also inhibit the expression of our own more genuine identity.

We have seen earlier that many children who are gender non-conforming *will not* become transgender adults. This tells us at least two things: one is that gender dysphoria at a prepubertal stage can eventually desist; the other, perhaps more important, is that during their development some children may need to explore and experience segments belonging to different genders. And this in turn means that gender identity is dynamic, not rigid, and open to changes over time.

9.7 Parental and Social Responsibilities

If what has been unravelled so far is persuasive, it follows that we have important moral reasons to rethink sex and gender, and how we, as adults, parents, carers or more generally citizens of a civilised society, can foster the flourishing of children's identities and contribute to the creation of a society that is able to embrace different sex and gender identities.

Dropping all sex and gender categories is not what I advocate here; it is doubtful whether such a move would be desirable or practicable. Yet, we have seen that there is no easy way to determine who is a female and who is a male (or who is a woman and who is a man). Most people *know* if they are a boy, a girl, or both or neither. Even young children will identify themselves as boys or girls (regardless of how they are registered at birth). How this awareness is acquired is not fully clear. However, it is unlikely that parents, school-teachers or meaningful others are able, at least on their own, to determine a child's sex and gender. In other words, there is no evidence that it is parents on their own who shape a child's gender.

Parents and significant others are instead able to inhibit (consciously or not) the expression of the child's true identity (as the case of *Smith v. Smith* illustrates). By the same coin they can foster a serene elaboration of gender identity, by being prepared to accept that there are various sexes and genders, that a child's gender identity may be different from what they might have expected, and that that identity may change over time. Parents thus have a moral responsibility to become aware of how damaging suppressing the experienced gender identity of a child may be both in the short and in the long term. Parents ought to acquire sufficient information about, say, appropriate nutrition and medical care, babies' hygiene and later education, and in a similar vein they have a responsibility to acquire sufficient information relating to the children's sex and gender development, in order to prevent or minimise the damaging effects of misguided stereotypes.

Parents cannot however be held uniquely responsible for fostering proper gender development. One area that deserves serious reconsideration is pre-school and school education. Research on gender stereotypes in education shows that a rigid gender divide still pervades nursery and primary education: children's stories are largely based on traditional definitions of gender roles, with strong gender clichés. Children's literature and traditional fairy tales usually feature women and men (transgender characters only appear for the first time in *Shrek*), women are often nurturant or endangered; men provide for them and protect them (Kolbenschlag 1979; Davies 1991; Pisetta 2004).

The Council of Europe has in 2014 reiterated its concern about this gender divide in compulsory education. It has stressed that education is a fundamental human right; as such it must respect people's equality and diversity. "Gender stereotyping presents a serious obstacle to the achievement of real gender equality and feeds into gender discrimination. Gender stereotypes are preconceived ideas whereby males and females are arbitrarily assigned characteristics and roles determined and limited by their sex. Sex stereotyping *can limit the development of the natural talents and abilities* of boys and girls, women and men, their educational and professional experiences *as well as life opportunities in general*" (Council of Europe 2014: 9) (my emphasis).

The State could (and should) also intervene to combat prejudice and stereotypes based on gender in a number of other ways. Cutas and I discussed some of these ways elsewhere (Cutas and Giordano 2013). These interventions could range from the abolition of clear gender markers, such as uniforms or gender specific schools, to the revision of children's literature so as to include narratives that portray gender and sex non-conformity as 'natural'. The introduction of facilities such as gender neutral toilets and gender neutral changing rooms both in schools and in public spaces can also temper people's perception of rigid gender and sex divides.

Of course any of these interventions in isolation is unlikely to suffice. And of course this is not to say that all schools that adopt school uniforms are inherently transphobic. Yet gender markers are likely to reinforce misguided stereotypes, and in this sense it may be important that these are also targeted. What policies may be implemented in order to improve people's understanding of the facts about sex and gender and draw the relevant ethical consequences that these facts bear also needs to be discussed; yet it is important to remember that gender discrimination should not be solely equated with acts of open violence or abuse (towards women for example, or gender minorities). Gender discrimination encompasses also the very blind acceptance of gender clichés and stereotypes. This is *per se*, even without open injustice or violence, a form of gender discrimination. Whereas obviously acts of injustice, abuse and violence have to be prevented and corrected, the gender stereotypes that provide the conceptual substrate to those acts also need to be eradicated.

The State has intervened in various countries with public policies or acts of parliament to combat intimidation (bullying) in schools, for example, or ethnic or religious discrimination; it has fostered public health with campaigns on smoking, obesity in children, HIV, and so on. Raising awareness about sex and gender should be a similar priority for the State in any civilized society.

9.8 Conclusions

Many of us would quite like the Stork to stay in business. The Stork is an innocent party of the many traditional tales that surround child birth. And many of us also undoubtedly enjoy the expectations that arise around the birth of a child, which include the preparation of clothes, cots, toys, and the mental anticipation of what life will be like, when this new person will be born. There is nothing inherently wrong with this: what is wrong is to ignore the facts and continue to remain anchored to ideas, such as that humankind is composed of males and females, which no longer have scientific support. If these ideas were harmless, perhaps we would not have a compelling reason to think again: but they can limit the children's freedom of expression and development, in the best case scenarios. In worse scenarios, they are responsible for the pathologisation of sex and gender non-conformity, and more worryingly, can trigger the violence, abuse, and constant humiliation of individuals, including children, who do not conform to the mainstream sex and gender divide.

We don't know whether a gender free society would be happier or more just. And gender clearly matters to people. What is clear, instead, is that rigid and polarised gender stereotypes are scientifically misguided and ethically problematic. Sex and gender should be treated as broad approximations; people can be placed in many locations along a broad and rich spectrum of possibilities. Sex and gender, moreover, are not always stable across the life of an individual: they can change over time. Without getting rid of the Stork, it may be time to renew her wardrobe. How much better to be able to pick amongst many colours, and how much relief for many families would be brought with the knowledge that their babies are just babies, and that different sex and gender identities are not illnesses, they are not disorders, but normal and natural ways in which a child may develop.

References

Archer, J., and B. Lloyd. 2002. *Sex and gender*. Cambridge: Cambridge University Press.

Bandura, A. 1965. Influence of model's reinforcement contingencies on the acquisition of imitative responses. *Journal of Personality and Social Psychology* 1: 589–595.

Baron-Cohen, S. 2003. *The essential difference: Men, women and the extreme male brain*. London: Penguin/Basic Books.

Bowlby, J. 1969. *Attachment and loss, Vol. 1, attachment*. Harmondsworth: Penguin.

Burke, S. 2014. *Coming of age, gender identity, sex hormones and the developing brain*, Doctoral thesis, Vrije Universiteit.

Burke, S., P. T. Cohen-Kettenis, D. Veltman, D. Klink, and J. Bakker. 2014. Hypothalamic response to the chemo-signal andro-stadienone in gender dysphoric children and adolescents. *Frontiers in Endocrinology*, May. doi:10.3389/fendo.2014.00060.

Bussey, K., and A. Bandura. 1992. Self-regulatory mechanisms governing gender development. *Child Development* 63: 1236–1250.

Connolly, P. 2003. *Transgendered peoples of Samoa, Tonga and India: Diversity of psychosocial challenges, coping, and styles of gender reassignment*. Presented at the Harry Benjamin International Gender Dysphoria Association Conference, Ghent, Belgium.

Council of Europe. 2014. *Gender equality strategy 2014–2017*, 9. Available at www.coe.int/t/dghl/standardsetting/equality/02_GenderEqualityProgramme/Council%20of%20Europe%20Gender%20Equality%20Strategy%202014-2017.pdf. Last accessed February 2016.

Cutas, D., and S. Giordano. 2013. Is it a boy or a girl? Who should (not) know children's sex and why? *Journal of Medical Ethics* 39: 374–377.

Davies, B. 1991. *Frogs and snails and feminist tales*. Sydney/London: Allen and Unwin.

Devor, H. 1989. *Gender blending: Confronting the limits of duality*. Bloomington: Indiana University Press.

Fausto-Sterling, A. 1993. The five sexes: Why male and female are not enough. *The Sciences* 33: 20–21.

Feinberg, L. 1996. *Trans gender warriors, making history from Joan of Arc to Dennis Rodman*. Boston: Beacon.

Giddens, A., and S. Griffiths. 2006. *Sociology*, 5th ed. Cambridge: Polity Press.

Gilligan, C. 1993. *In a different voice*, 6th ed. Cambridge, MA: Harvard University Press.

Goddings, A.L., K.L. Mills, L.S. Clasen, J.N. Giedd, R.M. Viner, and S.J. Blakemore. 2013. The influence of puberty on subcortical brain development. *NeuroImage* 69: 11–20.

Green, R. 2009. Gender identity disorder in children and adolescents. In *New Oxford textbook of psychiatry*, 2nd ed, ed. M.G. Gelder, N.C. Andreasen, J.J. Lopex-Ibor and J.R. Geddes. Oxford: Oxford University Press.

Grimbos, T., K. Dawood, R.P. Burriss, K.J. Zucker, and D.A. Puts. 2010. Sexual orientation and the second to fourth finger length ratio: A meta analysis in men and women. *Behavioural Neuroscience* 124: 278–287.

Gross, R. 2010. *Psychology*, 6th ed. London: Hodder Education.

Honekopp, J., and S. Watson. 2010. Meta-analysis of digit ratio 2D:4D shows greater sex difference in the right hand. *American Journal of Human Biology* 22: 619–630.

Ingalhalikar, M., A. Smith, D. Parker, T.D. Satterthwaite, M.A. Elliott, K. Ruparel, H. Hakonarson, R.E. Gur, R.C. Gur, and R. Verma. 2014. Sex differences in the structural connectome of the human brain. *Proceedings of the National Academy of Sciences* 111: 823–828.

Intersex Society of North America. 2016. *What is intersex?* San Francisco: Intersex Society of North America. Online at www.isna.org/faq/what_is_intersex. Accessed February 2016.

Kitzinger, C. 2001. Sexualities. In *Handbook of the psychology of women and gender*, ed. Rhoda K. Unger. New York: Wiley.

Kolbenschlag, M. 1979. *Kiss sleeping beauty good-bye: Breaking the spell of feminine myths and models*. Garden City: Doubleday.

Lamminmaki, A., M. Hines, T. Kuiri-Hanninen, L. Kilpelainen, L. Dunkel, and U. Sankilampi. 2012. Testosterone measured in infancy predicts subsequent sex-typed behavior in boys and in girls. *Hormonal Behavior* 61: 611–616.
Lewins, F. 1995. *Transsexualism in society. A sociology of male-to-female transsexuals*. Melbourne: Macmillan.
Lloyd, B., and G. Duveen. 1990. A semiotic analysis of the development of social representations of gender. In *Social representations and the development of knowledge*, ed. G. Duveen and B. Lloyd. Cambridge: Cambridge University Press.
Lorber, J. 1994. *Paradoxes of gender*. London: Yale University Press.
McIntyre, M.H., B.A. Cohen, and P.T. Ellison. 2006. Sex dimorphism in digital formulae of children. *American Journal of Physical Anthropology* 129: 143–150.
Meininger, E., and G. Remafedi. 2008. Gay, lesbian, bisexual and transgender adolescents. In *Adolescent health care: A practical guide*, ed. S.N. Lawrence, chapter 40. Minneapolis: Lippincott Williams and Wilkins.
Mill, J. S. 1869. *The subjection of women*. Online at http://www.constitution.org/jsm/women.htm. Accessed February 2016.
Nelson, R.J. 2005. *An introduction to behavioral endocrinology*, 3rd ed. Sunderland: Sinauer Associates.
Newbould, M. 2015. *Legal and ethical issues surrounding infantile and childhood genital surgery*. PhD thesis, University of Manchester.
Pisetta, A. 2004. *Genere e socializzazione scolastica. Una ricerca sulla rappresentazione dei modelli sessuali nei libri di testo per le elementari*. Tesi di Laurea, Università degli Studi di Trento, 2004. Online at www.tesionline.it/consult/indice.jsp?idt=11470. Accessed February 2016.
Ruspini, E. 2009. *Le Identità di Genere*. Milan: Carocci.
Scott, J.W. 1986. Gender: A useful category of historical analysis. *American Historical Review* 91(5): 1053–1075.
Skougard, E. 2012. The best interests of transgender children. *Utah Law Review* 2011: 1161–1201.
Steensma, T.D., R. Biemond, F. De Boer, and P.T. Cohen-Kettenis. 2011. Desisting and persisting gender dysphoria after childhood: A qualitative follow-up study. *Clinical Child Psychology and Psychiatry* 16: 499–516.
Valentine, S.E. 2009. Traditional advocacy for nontraditional youth: Rethinking best interest for the queer child. *Michigan State Law Review* 2008: 1053–1113.
van De Beek, C., S.H.M. van Goozen, J.K. Buitelaar, and P.T. Cohen-Kettenis. 2009. Prenatal sex hormones (maternal and amniotic fluid) and gender-related play behaviour in 13-month-old infants. *Archives of Sexual Behavior* 38: 6–15.
Vernon, R.A. 2009. Intersex and the molecular deconstruction of sex. *Journal of Lesbian and Gay Studies* 15: 267–283.
Vidal, I., D.B. Gorduza, E. Haraux, C.L. Gay, P. Chatelain, M. Nicolino, P.Y. Mure, and P. Mouriquand. 2010. Surgical options in disorders of sex development (DSD) with ambiguous genitalia. *Best Practice and Research. Clinical Endocrinology and Metabolism* 24(2): 311–324.
Wallien, M.S.C., R. Veenstra, B.P.C. Kreukels, and P.T. Cohen-Kettenis. 2010. Peer group status of gender dysphoric children: A sociometric study. *Archives of Sexual Behavior* 39(2): 553–560.
Warwick, I., E. Chase, and P. Aggleton. 2004. *Homophobia, sexual orientation and schools: A review and implications for action*, Research report No. 594, University of London. Online at webarchive.nationalarchives.gov.uk/20130401151715/http://www.education.gov.uk/publications/eOrderingDownload/RR594MIG1187.pdf. Accessed February 2016.
Will, J.A., P. Self, and N. Datan. 1976. Maternal behavior and perceived sex of infant. *American Journal of Orthopsychiatry* 46: 135–139.

Williams, T.J., M.E. Pepitone, S.E. Christensen, B.M. Cooke, A.D. Huberman, J.J. Breedlove, T.J. Breedlove, C.L. Jordan, and S.M. Breedlove. 2000. Finger-length ratios and sexual orientation. *Nature* 404: 455–456.
Zucker, K.J. 2001. Biological influences on psychosexual differentiation. In *Psychology of women and gender*, ed. R.K. Unger. New York: Wiley.

Legal Case

Smith v. Smith, No. 05 JE 42, 2007 WL 901599 (Ohio Ct. App. Mar. 23, 2007).

Chapter 10
The Lack of an Obligation to Select the Best Child: Silencing the Principle of Procreative Beneficence

Peter Herissone-Kelly

Abstract This chapter aims to show that prospective parents are not bound in their reproductive decision making by a principle of *procreative beneficence*. That is, they have no obligation (as Julian Savulescu, the principle's originator, famously thinks they *have*) to choose the possible child, from a range of possible children they might have, who is likely to lead the best life. I will summarise and clarify the content of previous papers of mine, in which I argue that since the sorts of considerations that underlie the principle of procreative beneficence do not constitute reasons when viewed from a perspective that it is fitting and appropriate for a prospective parent to take up, there can be no requirement for prospective parents to be moved by those considerations.

Keywords Principle of procreative beneficence • Savulescu • Dancy • Parenthood • Prospective parents

10.1 Introduction

Since 2006, when my paper "Procreative Beneficence and the Prospective Parent" (Herissone-Kelly 2006) was published, I have on a number of occasions revisited in print my claim that prospective parents are not bound by what Julian Savulescu calls *the principle of procreative beneficence* (Herissone-Kelly 2009, 2011, 2012). My position has been, then, that prospective parents can have no obligation at all to select, from a range of possible future children, the "best" child, understood as the one likely to lead the best life. In the course of the current chapter, I aim to do three things. I will offer a summary of my argument against Savulescu's principle,

P. Herissone-Kelly (✉)
School of Humanities and Social Sciences, University of Central Lancashire, Livesey House, Preston, UK
e-mail: PNHerissone-Kelly@uclan.ac.uk

K. Hens et al. (eds.), *Parental Responsibility in the Context of Neuroscience and Genetics*, International Library of Ethics, Law, and the New Medicine 69,
DOI 10.1007/978-3-319-42834-5_10

counter Savulescu's apparent objection to that argument and clarify some points about my position that have previously been, I must confess, rather muddier than they should have been. The summary will, being a summary, inevitably be informed by, and to a certain degree repeat, what I have said in my previous work. However, I hope to provide here some novel ways of explaining previously made points, and to bring together for the first time in one place a collection of observations that originally appeared in different papers.

I intend most especially by the end of the chapter to have brought out the significant extent to which my central argument against the principle of procreative beneficence relies on the truth of a claim made by Jonathan Dancy in his book *Moral Reasons* (Dancy 1992). Dancy thinks that considerations that function as reasons in one context will not necessarily function as reasons in all contexts in which they obtain, and thus that a thoroughgoing generalism about reasons is false. Furthermore—and crucially for my argument—Dancy's approach insists that when a consideration that operates as a reason in one context is rendered inoperative in another, it is not *overridden* by stronger reasons, but silenced. That is, it ceases to count as a reason altogether. It is my belief that this is true of the sorts of considerations that inform the principle of procreative beneficence. While in other contexts they function as reasons to act in such a way that we bring into existence people with the best lives, they need not so function when the decision-maker is a prospective parent, and the various possible people that could be brought into existence would be her children.[1] So, not only is there no *obligation* to act in accordance with Savulescu's principle, there is no *reason of any kind* to do so.

10.2 Savulescu's Principle

The principle of procreative beneficence, which I shall for the sake of ease call "PPB" from now on, is stated by Savulescu as follows:

> … couples (or single reproducers) should select the child, of the possible children they could have, who is expected to have the best life, or at least as good a life as the others, based on the relevant, available information. (Savulescu 2001)

In talking of selecting a child from a range of possible children, Savulescu has in mind a situation in which a number of embryos have been produced by In Vitro Fertilisation (IVF), and are subjected to pre-implantation genetic diagnosis (PGD) in order to establish their genetic constitutions. We should not let the use of the term "pre-implantation genetic *diagnosis*" mislead us here, though; Savulescu's picture is not a crude one of testing embryos for disease traits and

[1] When talking about prospective parents, I will throughout use the pronouns "she" and "her". This is only for the sake of ease, and to avoid the repeated use of unwieldy constructions such as "he or she" and "his or her".

discarding any that are found to be "unhealthy", though it does, at least in his hands, include an element of this. Rather, he thinks that, as our genetic knowledge and technologies become more advanced, we will be able to determine which of a range of embryos will be likely to develop into the child with the best life, even where no embryos in the range display disease traits. That is because we will be able to identify genetic factors that correlate with traits such as intelligence, athleticism, and impulse control, which Savulescu regards as indicators of the likelihood of a good life.

Although Savulescu holds that his principle identifies a moral obligation, he thinks it is one that only applies other things being equal. That is, it can on occasion be overridden by more pressing obligations, or by other features of the context in which a reproductive decision can be made. So, he writes, "Procreative Beneficence is a valid principle, albeit one which must be balanced against others" (Savulescu 2001). Elsewhere, and much more recently, he tells us that

> When I say moral obligation, I mean pro tanto moral reason, which must be weighed against other moral reasons to determine what we have most reason to do .... The obligation is a defeasible obligation or prima facie moral obligation. (Savulescu 2014)

This is simply a dismissal of rigorism in moral philosophy, and a recognition that very few of our obligations can be said to be absolute. However, where there are no countervailing obligations in place, PPB will invariably constitute what W.D. Ross would call our duty *sans phrase*, or our duty proper (Ross 2002). (Indeed, Savulescu's talk of a *prima facie* moral obligation is borrowed directly from Ross.) This notion that the obligation identified in PPB can only be avoided by being overridden by weightier moral reasons will be of relevance later, when we come to consider the way in which Savulescu is likely to object to my argument.

The core of Savulescu's case for PPB is remarkably straightforward, and as a consequence is likely at first sight to strike us as remarkably plausible too. Suppose we are faced with a choice about which of two embryos produced through IVF—let us call them A and B—to transfer to a womb for implantation. Let us take it as read that B has been shown through PGD to be what I call "the better-life embryo" (the embryo likely to develop into the person with the better life), and that A is therefore what I call "the worse-life embryo". Savulescu holds that we plainly have a reason to select B over A, and no reason to select A over B. What we have most reason to do in a particular situation is what we ought to do in that situation. So, again other things being equal, we ought to select B over A.[2]

[2] For a (justified) complaint that there is something implausible about the claim that we ought to do what he have most reason to do, see Hotke (2014). In what follows, I tacitly adopt Ben Saunders' "charitable" interpretation of Savulescu as holding that we morally ought to do what we have most *moral* reason to do. See Saunders (2014).

10.3 Two Objections to Savulescu's Principle

There are, of course, a number of objections to Savulescu's argument to be found in the literature. As a prelude to advancing my own objection, I want to consider just two of these.

One of the most cogent arguments against PPB is offered by Michael Parker (Parker 2007), who urges that the principle is simply unworkable. He plausibly maintains that what makes a human life go well, if indeed it can be specified at all, is likely to be too complex an affair ever to be tested for at the embryonic stage, no matter how sophisticated our technologies or our knowledge of genetics become. Pertinent though this objection is, it leaves PPB with a little too much life in it for my liking. That is, it allows the principle to be theoretically respectable, though contingently inapplicable: Parker seems to want to say, or at least to leave it open to say, that if prospective parents *were* in a position to tell that, out of A and B, B would be likely to develop into the person with the better life, then they *would* be morally required to select B. I, on the other hand, want to maintain that even if it were possible to rank a number of possible lives from likely best to likely worst, still the selection of the child likely to lead the best life would not be even *prima facie* obligatory for prospective parents. In fact, I argue that, so long as all the other possible children in the available range would be likely to lead lives worth living, there is not even a *reason*, never mind any sort of obligation, for a prospective parent to choose any one child over any other.

This last claim is one with which another opponent of PPB, Rebecca Bennett (Bennett 2009), would agree. Nonetheless, I think that Bennett's own objection to the principle is unsuccessful, though a consideration of it and the ways in which it fails will pave the way for my own criticism of PPB. A useful way into Bennett's position is to consider the differences between the two following scenarios.

10.3.1 Scenario 1

A is an IVF embryo which, if transferred to a womb, will eventually develop into a person A*. The prospective parents of A* have access to a risk-free and affordable intervention which, if it were carried out on A, would give A* a considerably better life than she would have had in the absence of the intervention. If they choose not to make use of the intervention, it would seem plausible to say that A* has cause for complaint. She has, that is, apparently been wronged by her parents, who could easily have benefited her, but neglected to. Had they arranged for the intervention to be carried out on A, A* would have been benefited. It is therefore easy to make out a case that the prospective parents have a moral obligation to carry out the intervention.

10.3.2 Scenario 2

A is an IVF embryo which, if transferred to a womb, will eventually develop into a person A*. B is a different IVF embryo which has, nonetheless, been produced from a sperm and egg from the same two prospective parents as A. If transferred to a womb, B would eventually develop into a person B*. It is highly likely that B* would enjoy a considerably better life than would A*, though both would lead lives worth living. B is therefore what I call the better-life embryo, and A the worse-life embryo. Now suppose that the prospective parents select embryo A, and thus that A* comes into existence. A*, it would seem, has from her own perspective no cause for complaint; she has no reason to wish that B had been selected instead, just because, had B been selected, A* would not have been benefited. She would not have lived a better life; instead, a completely different person, B*, would have lived a better life than A*'s. This fact is, so to speak, nothing to A*. She has not been wronged by being brought into a worthwhile existence in place of somebody else with a still more worthwhile existence. Indeed, in so far as being brought into a worthwhile existence is a benefit, she has been benefited. Nor has B* been wronged by not being brought into existence, since it is impossible to wrong someone who does not exist. (Of course, B* would have been benefited by being brought into existence, but, as we have seen, the same is true of A*.)

Bennett's point is that because no particular individual is harmed or wronged when, in Scenario 2, A is selected over B, there can be no moral obligation to select B over A (Bennett 2009). The thought, then, is that there can only be a moral obligation to refrain from a certain course of action when (as in Scenario 1) that course of action would harm a particular individual or particular set of individuals, making them worse off than they otherwise would have been. Conversely, there can only be a moral obligation to carry out a certain course of action when (as, again, in Scenario 1) that course of action would benefit a particular individual or particular set of individuals. In Scenario 2, neither A* nor B* would be harmed, whichever of A or B is selected. And arguably, in so far as being brought into worthwhile existence is a benefit, then while B* would be benefited by B's being selected, A* would equally be benefited by A's being selected. There can therefore be no moral obligation to choose either B over A, or A over B.

Now, Savulescu agrees that the child who would develop from a worse-life embryo such as A would not be harmed by A's being selected. This fact, he thinks, exempts PPB from being legitimately enforceable through legislation, since the principle of reproductive autonomy leaves any non-harmful procreative decision in the lap of prospective parents. But he maintains that there is still a wrong involved in choosing A, and thus that there can be wrongs that are non-person-affecting, or that harm no particular individual or set of individuals. That this is the case is ostensibly shown by certain compelling thought experiments, most notably those produced by the philosopher Derek Parfit in his landmark book *Reasons and Persons* (Parfit 1984). Those thought experiments describe scenarios in which a choice is offered between a course of action that will result in the existence of a particular set of individuals S_1, or a different course of action that will result in the existence of a distinct set of individuals S_2, where the members of S_2 would possess a quality of life significantly lower than that of the members of S_1, but not so low as to be anywhere near intolerable. The intuition appealed to here is that we plainly have a moral obligation to adopt the course of action that will lead to the existence of S_1, even though nobody would be harmed by our choosing the course of action that would result in the existence of S_2.

Savulescu offers his own example of just such a Parfitian thought experiment:

> A poor country does not have enough power to provide power to its citizens during an extremely cold winter. The government decides to open an old and unsafe nuclear reactor. Ample light and heating are then available. Citizens stay up later, and enjoy their lives much more. Several months later, the nuclear reactor melts down and large amounts of radiation are released into the environment. The only effect is that a large number of children are born with predispositions to early childhood malignancy.
>
> The supply of heating and light has changed the lifestyle of this population. As a result of this change in lifestyle, people have conceived children at different times than they would have if there had been no heat or light, and their parents went to bed earlier. Thus, the children born after the nuclear accident would not have existed if the government had not switched to nuclear power. They have not been harmed by the switch to nuclear power and the subsequent accident (unless their lives are so bad they are worse than death). (Savulescu 2001)

Two interrelated points, thinks Savulescu, can be drawn from the "Nuclear Accident" example. First, most of us will judge that the government acted wrongly in choosing to open the unsafe nuclear reactor. Secondly, in making such a judgement, we are implicitly appealing to a notion of non-person-affecting wrong. As Savulescu puts it, "we must claim that a wrong was done, but no one was harmed. We must appeal to something like the Principle of Procreative Beneficence" (Savulescu 2001).[3]

Bennett, however, declares herself unmoved by such thought experiments, and by the intuitions to which they typically give rise in those who encounter them. These, she thinks, are not genuine moral intuitions, but expressions of preferences only: we would *prefer* a world inhabited by the members of S_1 rather than by the members of S_2; we would *prefer* a world in which there had been a happier set of individuals and no nuclear accident, rather than a different and less happy set of individuals as a result of that accident. But we could have no moral reason to produce that world, just because no-one would be benefited by our doing so, or harmed by our failing to do so. In the same way, there is no specifically moral reason to

[3] It seems to me, incidentally, that there is an unremarked flaw in Savulescu's example, though not one that damages the point he wishes to make about the existence of non-person-affecting wrong. That is, there presumably *is* a particular set of individuals whose lives are made worse by the occurrence of the accident: namely, the individuals alive at the time that it happened, in so far as they were subject to much more worry and concern than they would have been had the unsafe reactor not been opened. I am not sure that that level of worry could realistically be compensated for by their having previously had greater opportunity to stay up late, or by the novelty of having sex later in the evening.

select embryo B over embryo A. How could there be, when no matter which embryo we choose, there is no child for whom our action is worse?

Though at first sight there might seem to be something compelling about Bennett's claim, I think the strength of our intuitions in the Parfitian cases ought to give us pause. These, after all, are not simply strong intuitions that a world inhabited by people with better lives would be something preferable, but intuitions that it is something we *ought* to prefer, something we have moral reason to produce if we are in a position to do so. This might make us suspicious of Bennett's argument. Our suspicion will, I submit, only be increased when we come to subject her argument to closer scrutiny. That is, it strikes me that careful attention to what Bennett has to say will reveal a certain (no doubt unintentional) sleight-of-hand. In short, it seems to me that, stripped down to its essential points, her argument can be set out as follows:

(1) There are two sorts of putative moral wrongs: person-affecting moral wrongs, and non-person-affecting wrongs;
(2) The putative non-person-affecting wrongs are not person-affecting;
(3) Therefore, the putative non-person-affecting wrongs are not really wrongs at all.

It should be clear when it is set out in this way that Bennett's argument is question-begging: we can only reach the conclusion that supposed non-person-affecting wrongs are not really wrongs at all, if we tacitly assume that very conclusion as an extra premise,

(1′) Putative non-person-affecting wrongs are not really wrongs at all.

or if we accept as a premise a corollary of that conclusion:

(1″) The only wrongs there are, are person-affecting wrongs.

Bennett's argument for her conclusion, then, fails. That, of course, does not entail the falsity of the conclusion. Nonetheless, I think it can be independently shown that there are indeed non-person-affecting wrongs.

10.4 Two Perspectives on the Lives of Possible Future Persons

We might start a defence of the concept of non-person-affecting wrongs with a reasonably straightforward observation. In the previous section, while discussing Scenario 2, I suggested that, where A is a worse-life embryo that would develop into a person A* and B a better-life embryo that would develop into a person B*, it is equally the case, where A* and B* both have worthwhile lives, that A* will be benefited by A's being selected, and B* will be benefited by B's being selected. And I suggested that an upshot of this might be that there is no moral reason to select B over A. But is that really true? It is possible to make a nuanced but significant distinction here, by pointing out the difference between the following two propositions:

(1) It is equally the case that A* and B* will be benefited by their respective embryos being selected.
(2) It is the case that A* and B* will be equally benefited by their respective embryos being selected.[4]

Proposition (1) is true, and its truth clearly does not, considered by itself, give us even a *prima facie* moral reason to select B over A. A similar claim could be made about proposition (2) if it were also true, as for instance it would be if it were entailed by the true proposition (1). However, proposition (2) is neither entailed by proposition (1), nor is it independently true. A* and B* will not be equally benefited by being brought into existence: B* will be more benefited by B being selected than A* will be benefited by A being selected. In other words, although it would not be better for A* if B were to be selected—since in that case A* would not exist—it is nonetheless the case that B*'s life would be better for B* than A*'s life would be for A*. And that, we might be tempted to think, *does* give us moral reason to select B over A. It seems to me that it is worth investigating and developing this thought, in order to shed light on the notion of non-person-affecting wrong.

Bennett's dismissal of the concept of non-person-affecting wrong is seemingly predicated on the belief that the only morally relevant way to think about the lives of possible future people is to do so from what I call the internal perspective. Although we are also able to think of the lives of possible future people from an external perspective, a perspective that compares the quality of one possible life with another, such a perspective, Bennett would say, is unable to ground any genuine moral obligations. I think that this is not the case, and thus that there can be moral obligations rooted in the external perspective just as there can be duties that are anchored in the internal perspective. This might appear to be good news for Savulescu and those who agree with him, just because PPB is an external-perspective principle. Nonetheless, I will go on to argue that when prospective parents are considering the lives of their possible future children, it is not only permissible, but fitting and appropriate for them to do so only from the internal perspective, and that they are therefore not bound by any reasons, moral or otherwise, that show up only when the external perspective is adopted. If what I have to say here is to be fully understood, I need first to explain the difference between the internal and the external perspective.

In thinking about the life of a possible person A*, we adopt the internal perspective when we (i) "imaginatively inhabit" that life, imagining what it would be like to live it, and (ii) make the sort of "better" and "worse" judgements that we would make about A*'s life *if we were A**. All the "better-for" or "worse-for" judgements made from the internal perspective will be judgements about what will be better for

[4] If it is difficult to see how these two propositions differ in content, compare the two claims "It is equally the case that Michael and Gertrude are tall", and "It is the case that Michael and Gertrude are equally tall". The former may be true while the latter is false.

or worse for *A**. When we adopt the internal perspective, we remain unmoved by the talk of better and worse lives that features in principles like Savulescu's: such talk belongs only to, and has any purchase only from, the external perspective.

In thinking about the life of a possible person A* from the *external* perspective we (i) imaginatively inhabit that life, attempting to gain a sense of what it would be like to live A*'s life; (ii) imaginatively inhabit the life of another possible person B*, attempting to gain a sense of what it would be like to live B*'s life; and (iii) *draw back from the perspective of both A* and B** in order to make a judgement about which life is better. This involves judging whether B*'s interests would be better met by B*'s life than A*'s interests would be met by A*'s life. The perspective from which the judgement is made here is "external", because it stands outside either life, and does not involve any attempt to determine what would be better for either A* or B*.

Bennett's claim is that any principle (including PPB) that is grounded in the external perspective cannot identify a genuine moral obligation, just because no particular individual or particular set of individuals could gain better lives than they otherwise would have had from that principle's being acted upon. We have already seen that the argument that Bennett provides for this position fails as a result of being question-begging. Our task now is to demonstrate the independent falsity of Bennett's conclusion.

Bennett's assumption seems to be that a concern with welfare can only be a concern with the welfare of particular individuals, and that since morality is centrally concerned with welfare, its business can only be with the benefiting of, or the prevention of harm towards, particular individuals. However, it seems to me to be clearly the case that external perspective principles are authentically concerned with welfare too: they are concerned with bringing about that set of individuals whose interests are best met by their lives, beings with greater rather than lesser amounts of welfare. Indeed, Savulescu and Guy Kahane urge that it is potentially misleading to characterise action on principles like PPB as straightforwardly non-person-affecting. PPB can, they claim, be expressed in a "wide person-affecting form" as well as in an "impersonal form" (Savulescu and Kahane 2009).

Given their concern with welfare, then, external perspective principles are all at least candidates for genuine moral obligations; there is nothing that *a priori* rules out their being so regarded. The fact that there can be duties to bring about a set of individuals S_1 whose lives are better for them than the lives of the members of an alternative set S_2 would be for those members can explain the strong intuitions called forth by examples such as Savulescu's Nuclear Accident scenario. We want to say that *of course* the government in that example ought not to open the unsafe nuclear reactor, just because doing so risks a significant negative impact on welfare.

It is my view, then, that there can be external perspective principles that express genuine moral obligations. Nonetheless, I do not believe that PPB is to be numbered amongst them. This commits me to the belief that there is no one, invariant answer to the question of which perspective, internal or external, should be adopted when thinking about the lives of possible future people. It also commits me to the belief that prospective parents at the very least *need not* take up the external perspective.

10.5 Prospective Parenthood and the Internal Perspective

Governmental policy makers, of the sort that feature in Savulescu's Nuclear Accident example, arguably have a duty to make the societies they govern into the best societies they can. This will in part involve their selecting policies that ensure the interests of those existing people they govern are met as well as they can be. But it may also on occasion involve their making decisions that affect which of (for example) two sets of people come into existence. It is appropriate—indeed, I would urge that it is obligatory—for policy makers to take up the external perspective here, and to reason from within that perspective; they should positively eschew the internal perspective.

Nonetheless, the heart of my argument against PPB is that prospective parents thinking about which possible future child to bring into existence have no such obligation to adopt the external perspective, nor (*a fortiori*) to be moved by the considerations that show up as reasons from that perspective. That is because, for a prospective parent in such a situation, the internal perspective is an entirely appropriate and fitting one to adopt.

The thought here is this: in order to make a decision about which of A (the worse-life embryo) and B (the better-life embryo) to select for transfer to a womb, the prospective parent must imaginatively inhabit both possible future lives: the life that A* is likely to live, and the life that B* is likely to live. In doing this, she is imagining, in both cases, the lives likely to be led, not as it were by *just any person*, but by *her own possible offspring*. It is therefore appropriate that she should relate to those potential children's lives in a way that is proper for a parent. Doing so will involve her in an extremely close identification with the possible subject of each life she imagines—so close that it will include and give weight to the sorts of assessments that each child would be likely to make of his own life: the sorts of assessments portrayed, indeed, in Scenario 1 earlier in this chapter. In short, then, the standpoint in question is one that includes just those features from which the external perspective abstracts; it is the internal perspective. The only judgements about welfare accessible from this perspective will concern what would be better for A* or better for B*. The transpersonal welfare judgements of the external perspective will find no foothold here. As a result, the consideration that B*'s life would be better than A*'s (and so, better for B* than A*'s life would be for A*), while it may be acknowledged, will not show up as any sort of reason to select B* over A*.[5]

[5] The rejection of PPB does not entail that there are no situations in which a better-life embryo B ought to be selected over a worse-life embryo A. Suppose, for example, that A*'s life would be likely to be of such poor quality that it would not be worth living. In such a case, A ought not to be selected. But the reason for choosing B over A in such a situation would not be that B*'s life would

In a recent paper, Ben Saunders has drawn attention to a certain prevarication in my past work on the topic of PPB. I have said that the internal perspective is an appropriate one for prospective parents to take up when thinking about the lives of their possible future children. Does that mean that I think it is impermissible for them to take up the external perspective in such situations? Or does it mean simply that either perspective is permissible (and thus, while PPB does not represent an obligation for prospective parents, neither would it be morally impermissible for them to be guided by it)? Saunders comments that "[Herissone-Kelly's] remarks suggest a certain (studied?) ambivalence between these two versions of the argument" (Saunders 2014). While I can report that the ambivalence was not studied, I do concede, with hindsight, that it was genuinely present. That being the case, I would like briefly to give an unambiguous statement of my considered position on this question.

In saying that it is "appropriate" for prospective parents to take up the internal perspective when faced with the option of which possible child to select, I mean to say something a little stronger than that it is merely permissible for them to do so, but something definitely weaker than that it is obligatory.

An analogy from the realm of etiquette might help explain what I mean here. Suppose that a novice concert-goer attends a classical music event with a knowledgeable companion. He notices that nobody applauds at the end of movements and asks his companion if nevertheless it would be appropriate to applaud at this point. She tells him that it would not. What he has asked, of course, is whether by the rules of concert etiquette it is *permissible* to applaud at the end of a movement. Now suppose that the piece ends and our inexperienced concert-goer notices that everybody else claps enthusiastically, but does not himself join in. His companion tells him that it is now appropriate to clap. He acknowledges this new information, but says that he has decided not to show appreciation of the performance, despite having enjoyed it greatly. In so doing, he has interpreted her claim to be that it would be merely permissible to applaud now, and so something one can equally choose not to do if one pleases. This, it seems to me, is a misunderstanding of the situation on his part, and of what his companion means when she mentions what it would be appropriate to do. To be sure, she is not suggesting that it is, according to the rules of concert etiquette, *obligatory* to applaud. And yet she is not saying either that it is merely permissible. Instead, her claim is that it would be the *proper* thing to do, the sort of thing to be expected. Although he will not be ejected from the concert hall for not clapping, he might expect to be looked at askance by other audience members, or by the conductor.

In a similar way, in saying that it would be appropriate for a prospective parent to adopt the internal perspective in thinking about the lives likely to be lived by her possible future children, I am saying this is something that might be expected from a good prospective parent: that it is implied by the normative character of her role. And yet I now unequivocally want to stop short of saying that it is in any sense

be better than A*'s. Instead, it would simply be that B* would be the only one of the two possible future persons whose life would be worth living.

obligatory; there has been no dereliction of duty if the prospective parent takes up the external perspective. We might want to say that she simply has not been as parental as she might have been, where the notion of being parental is a normatively charged one.

What it is important to note is that, on my account, it is certainly permissible (although it is also something more) for prospective parents to adopt the internal perspective. That being the case, it is not obligatory for them to assume the external perspective, and therefore not obligatory for them to count any considerations that show up only from that perspective as reasons for action. PPB does not identify a genuine moral requirement for prospective parents.

10.6 Silencing (Not Overriding) PPB

What might Savulescu say in response to this position? Although he has cited my work a number of times, he has never directly addressed my argument in print. In person, though, his objection has been this. Where B*'s life would be likely to be better than A*'s there is a reason to select B over A, and no countervailing reason to select A over B. Where there is no countervailing reason, the only reason that there is ought to move us to action. This is just a fact about practical rationality.

My response is that it may well be a fact that where there is a reason to φ (where φ-ing is a type of action), and no countervailing or overriding reason not to φ, then one ought to φ. But to produce this fact as a way of defeating my position is straightforwardly to misunderstand what it is that I am arguing. I do not need to provide a countervailing reason to override the supposed reason that a prospective parent has to select B over A. After all, it is not my belief that the prospective parent has a reason for not choosing one embryo over the other. Rather, my claim is that she has no reason for choosing one over the other. This is a delicate distinction, but a crucial one. My thought is that there is something about the internal perspective—which, again, is a fitting perspective for a parent to adopt—which does not *override* the sorts of reasons visible from the external perspective, but *silences* them. That considerations that function as reasons in one situation can fail to function as any sort of reason at all in other situations—can be silenced rather than overridden in those situations—is a central claim of the moral particularism advanced by Jonathan Dancy.

The intuitively compelling nature of this claim is, I think, powerfully brought out by an example supplied by Dancy. He recalls hearing the moral philosopher R.M. Hare claim that cruel, pain-inflicting actions from which their perpetrators gain satisfaction have at least this to be said for them: they give pleasure to those who carry them out (Dancy 1992). Hare's thought seems to be that the pleasure to be gained from the performance of such actions is *some* sort of consideration in favour of their being performed, albeit one that is massively outweighed by the reasons against. Dancy, I think rightly, finds this claim to be quite startlingly wrong. He wants to maintain that, in the case of such acts, the fact that they give pleasure to their perpetrators is no reason at all—not even an invariably overridden reason—in their favour. Although the fact that an act will give pleasure to the person who performs it is in many contexts a reason why that act should be performed, in this sort of context it is silenced (rather than overridden) by the presence of other considerations that function as reasons against performing the act.

Now, the sort of silencing outlined in this example involves a consideration that can function as a reason *for* action in some contexts, being silenced by other considerations that function as reasons *against* action. But this, Dancy thinks, need not be the only way in which silencing occurs. Considerations can also be silenced by background conditions that do not themselves function as reasons, or motivators. This is precisely, I submit, what happens when the person required to choose which of two possible children to bring into existence, where one will be likely to have a better life than the other, is the prospective parent of both and adopts the internal perspective. There is not some further reason for action, peculiar to the context, that silences what we might call the default obligation to bring into existence the person who will have the better life. Rather, that default obligation is silenced by the background conditions of the case, in the shape of the identity of the person called upon to do the selecting. That B* will be likely to have a better life than A* provides no reason at all for a prospective parent to choose B over A. Nor is there any reason to select A over B.

If there is no reason for the prospective parent to select a better-life embryo over a worse-life embryo, then there is no *moral* reason to make such a selection. And if that is the case, then PPB cannot represent a moral obligation (since we can only have a moral obligation to do what we have moral reason to do). The sorts of considerations on which PPB is founded, and which function as moral reasons in other situations, are silenced in a context in which a prospective parent envisages the lives of her possible future children, as she can quite fittingly do, from the internal perspective.

References

Bennett, R. 2009. The fallacy of the principle of procreative beneficence. *Bioethics* 23(5): 265–273.

Dancy, J. 1992. *Moral reasons*. Oxford: Blackwell.

Herissone-Kelly, P. 2006. Procreative beneficence and the prospective parent. *Journal of Medical Ethics* 32(3): 166–169.

Herissone-Kelly, P. 2009. Two varieties of "better-for" judgements. In *Harming future persons: Ethics, genetics and the non-identity problem*, ed. M. Roberts and D.T. Wasserman, 249–263. Heidelberg: Springer.

Herissone-Kelly, P. 2011. Reasons, rationalities and procreative beneficence: Need Häyry stand politely by while Savulescu and Herissone-Kelly disagree? *Cambridge Quarterly of Healthcare Ethics* 20(2): 258–267.

Herissone-Kelly, P. 2012. Wrongs, preferences, and the selection of children: A critique of Rebecca Bennett's argument against the principle of procreative beneficence. *Bioethics* 26(8): 447–454.

Hotke, A. 2014. The principle of procreative beneficence: Old arguments and a new challenge. *Bioethics* 28(5): 255–262.
Parfit, D. 1984. *Reasons and persons*. Oxford: Oxford University Press.
Parker, M. 2007. The best possible child. *Journal of Medical Ethics* 33(5): 279–283.
Ross, W.D. 2002. *The right and the good*. Oxford: Clarendon.
Saunders, B. 2014. Is procreative beneficence obligatory? *Journal of Medical Ethics* 41(2): 175–178.
Savulescu, J. 2001. Procreative beneficence: Why we should select the best children. *Bioethics* 15(5/6): 413–426.
Savulescu, J. 2014. The nature of the moral obligation to select the best children. In *The future of bioethics: International dialogues*, ed. A. Akabayashi, 170–182. Oxford: Oxford University Press.
Savulescu, J., and G. Kahane. 2009. The moral obligation to create children with the best chance of the best life. *Bioethics* 23(5): 274–290.

Chapter 11
Parental Responsibility and the Principle of Procreative Beneficence in Light of Assisted Reproductive Technologies

Francisco Güell Pelayo

Abstract The aim of this article is to investigate whether techniques used for artificial reproduction safeguard or promote the future child's welfare, and whether they are capable of offering potential future children the best possible chance for the best life, in keeping with guidelines derived from the Principle of Procreative Beneficence. This analysis will be important for discerning the parental responsibility of couples or single reproducers who plan to use or have used any of these techniques, and also for those who defend the Principle of Procreative Beneficence, which implicitly entails the use of techniques of assisted human reproduction. The paper concludes that prospective parents should be informed not only of the specific level of risk and potential damages associated with each IVF technique, but also of the fact that given the available evidence none of the standard IVF techniques can be considered to be risk free, there is reason to believe that none of these techniques can be reconciled with the responsibility of prospective parents to the promote welfare of future children and/or to offer them the best possible life.

Keywords Principle of procreative beneficence • Assisted reproductive technologies (ART) • Epigenetics • Embryo • Gametes • In Vitro Fertilization (IVF)

F. Güell Pelayo (✉)
Mind-Brain Group, Institute for Culture and Society (ICS), Universidad de Navarra, Office 2160, Edificio Bibliotecas (ICS) – Campus Universitario, 31009 Pamplona, Spain
e-mail: fguell@unav.es

K. Hens et al. (eds.), *Parental Responsibility in the Context of Neuroscience and Genetics*, International Library of Ethics, Law, and the New Medicine 69,
DOI 10.1007/978-3-319-42834-5_11

11.1 Introduction[1]

Parental responsibility encompasses the rights, duties, powers, and authority that parents have regarding their children. The Medical Protection Society states that parents must behave dutifully towards their children, and should do what is reasonable in all circumstances to safeguard and promote their children's welfare (Medical Protection Society 2012). Advances in our knowledge about epigenetics over the last decade have made it possible to extend this concern to the protection and welfare of children even before they are born. Indeed, as Hens argues elsewhere in this volume, the behaviour and habits of prospective parents and their potential impact on the future health of the child are increasingly considered as issues of parental responsibility alongside environmental exposure in utero and the behaviour of future parents prior to conception (Hens in this book). For example, as also Dondorp and de Wert point out in their chapter, prenatal child protection and the punishment of expecting mothers for the use of alcohol and other drugs during pregnancy are currently given serious consideration (Olsen 2015, Dondorp and de Wert in this book).

Initially, one might think that this expansion of parental responsibility is driven by the same concerns that first led to the formulation of the *Principle of Procreative Beneficence* (PPB), presented by Savulescu for the first time in 2001 (Savulescu 2001) and later modified in 2009 (Savulescu and Kahane 2009). The PPB states that parents have a moral obligation to use available reproductive technologies to create children who have the best chance of enjoying the best possible life. Following this guideline, Savulescu and Kahane maintain that "there is reason to obtain and use all genetic and other information about disease susceptibility and non-disease states to make a decision to select the most advantaged child" (Savulescu and Kahane 2009). Moreover, as they point out, in vitro fertilization (IVF) is the relevant technological context in which these matters are considered and, as a consequence, following the PPB may require the use of invasive technology at pre-conception (to obtain, modify and/or select gametes), post-conception (to create and select embryos), and prenatal stages in order to create children with the best chance of the best life (Savulescu and Kahane 2009). For the purposes of this chapter, it should be noted that the PPB does not explicitly state that it is specifically embryos that should be selected; rather, it states that the "most advantaged children" should be selected, and here this statement is taken to imply that selection can take place at the level of embryos but also at the level of gametes.

The aim of this paper is to investigate whether IVF is indeed able to safeguard or promote the welfare of children created in this way, as required by parental responsibility, and whether it is capable of offering children the best chance of the best life, as required by the stricter guidelines of the PPB. This analysis is important for defining the parental responsibility of couples or single individuals who are think-

[1] I am especially grateful to Nathaniel Barrett for his generosity. His numerous suggestions related to the ethical argumentation have helped make the paper much stronger. I am also grateful to the Institute for Culture and Society (Universidad de Navarra) and all the "Mind-Brain Group" colleagues. This work was co-funded by Obra Social La Caixa.

ing of using IVF to become parents. It is also important for the defenders of the PPB, given that this entails the use of IVF and embryo and/or gamete selection (Savulescu and Kahane 2009).

For ease of comparison with other approaches, in the following discussion I refer occasionally to a scenario employed by Herissone-Kelly in this book. In his *Scenario 1*, Herissone-Kelly states that if we have an IVF embryo A, the prospective parents will have a moral obligation to carry out an intervention if A would benefit (becoming A* after the intervention). To clarify questions of parental responsibility and the PPB, I enrich this scenario as follows. First, if the prospective parents of Herrissone-Kelly's scenario have a moral obligation to carry out an intervention if A would benefit as A*, I propose that parents have a similar moral obligation to avoid those interventions that may pose a danger for A* that outweighs the expected benefit. Second, following Savulescu's argument (Savulescu and Kahane 2009), I suggest to bring the gametes [Gs] into consideration. As mentioned above, Savulescu indicates the possibility of pre-conception and pre-implantation selection in conformity with the PPB, and this includes the selection of both embryos and gametes for the creation of the most advantaged child (ibid.). Gs are ontologically different entities than A, but because Gs have a direct influence on A they are relevant to the present investigation. Thus the scenario to be considered here involves Gs, A and A*, and my purpose will be to analyse how IVF and related techniques affect Gs and A, and, in turn, how these interventions can affect A*.

In cases where there is more than one embryo (A1, A2, A3...) whatever is stated in the first part of the scenario [how IVF techniques affect Gs and A] will be valid for every A(n) that exists prior to selection. For present purposes, the only notable difference between a single A and a series of A(n), is that the latter case may involve the performance of pre-implantation genetic diagnosis (PGD), which constitutes an additional type of intervention with possible impact on A and thus A*. In short, for the purpose of analysing the effects of IVF techniques on the welfare of future children, a series of embryos only differs from single embryos in respect of the additional influence of PGD, provided this technique is used.

For these reasons, I have structured this chapter as follows. In the first section I provide a brief description of the development of gametes and the pre-implantation embryo, paying special attention to the epigenetic dimension and its role in cellular differentiation. This description is necessary for properly specifying and explaining the relevant phases and techniques of intervention as well as for considering the relevant issues of responsibility. In section two, I explain the possible impact that each phase/technique can have on gametes and embryo development, and in the third section I review research that is relevant to analysing the health of persons born as a result of IVF. The fourth section provides a discussion of various issues of responsibility that arise from the previous analysis and, in the fifth section, I consider the central questions of whether IVF and related assisted reproductive technologies (ARTs) can be seen as safeguarding or promoting the child's welfare, and whether they can offer our children the best chance for the best life, in keeping with the PPB.

11.2 Scientific Background

11.2.1 The Developmental Dimension

The purpose of this section is to provide the relevant scientific background for a consideration of IVF techniques, focusing on the genetic and epigenetic dimensions of development and the way in which this dynamic process affects the maturation of the gametes, the formation of the zygote, and the proliferation and differentiation of embryonic cells.

Every cell (or blastomere) that makes up the embryo has the same sequence of nucleotides, or DNA. However, in this context it is important to distinguish between the sequence of nucleotides and the biochemical and structural configuration of the chromatin fibre of DNA, what is known as the "epigenome" (Güell 2013). The crucial point is that while the DNA *sequence* is the same in all the cells of an organism, the *configuration* of the DNA is different. Epigenesis refers specifically to alterations of the DNA configuration by means of changes to chemical signals (commonly called "epigenetic marks") inherited from the cell's lineage or previous developmental stage. The precise configuration of the genetic material in each of the cells in the embryo determines gene expression, and thus differences of configuration are crucial to explaining the differentiation of cells or tissues. However, it is important to note that gene expression also depends on the composition of the cell cytoplasm—in other words, on the available cell machinery—as well as on factors of cellular and nuclear composition and configuration which initiate and regulate the process of gene expression. Some of this machinery is incorporated from the environment of the cell, and depending on the developmental stage, this environmental contribution may be fundamental to the correct development of the embryo.

Another important point is that during gametogenesis the DNA of the ovum and the spermatozoon acquire certain peculiarities at the epigenetic and cytoplasmic levels, that are carried over by the zygote and all descendent cells of the organism. Because of the parental origin of this differential epigenetic configuration, this phenomenon has been called "parental imprinting". As a result of parental imprinting, the DNA configuration pattern or, more concretely, the methylation pattern, in the zygote and in all subsequent cells of an organism differs depending on its origin (ovular or spermatic). These biochemical modifications of the gametes are crucial factors for differential gene expression during the development of the embryo. Furthermore, after conception, additional modifications to the methylation pattern occur over the entire course of development, contributing to the differentiation of the more than 120 cell lines that make up the diverse tissues of the adult organism.

In summary, precise variations of methylation patterns during gametogenesis are essential for conception, while further changes of genetic configuration and cytoplasmic load that occur during the peri-conceptional period are essential for the differentiation of somatic cells. These epigenetic details are worth noting because they show that the healthy development of an embryo (let alone the best chance for the best possible life) depends on much more than having "good genes."

The importance of this last point cannot be stressed enough—it signals a radical shift in our scientific understanding of development with major ramifications for our understanding of the ethics of assisted reproduction. Of course, because discussions of the ethics of IVF intervention rarely enter into the details of development, it is hard to say exactly how this scientific background figures in ethical reasoning. Nevertheless, it is more likely than not that a heavily gene-centred view of development is taken for granted in most discussions, because this has been the dominant scientific view for most of the past half-century. Scientists have long described DNA as the "blueprint" for a person, and when taken literally this suggests that gametes are like envelopes, each containing half the total information needed for a complete "blueprint." When this is the guiding image for our reflections on IVF intervention, we are much more likely to be concerned about the genetic information inside the envelope than the envelope itself. Any intervention that alters this information would be, according to the "blueprint" metaphor, a potential threat to the "informational essence" of a future individual.

Now, the significance of the revised scientific background just summarized above is that it reveals the falsehood of the "blueprint" image: what we formerly treated as merely an envelope for the delivery of essential genetic information is, in fact, an essential part of how genetic information is arranged and expressed as such. Thus to tamper with the "envelope"—altering cytoplasmic load and other factors related to gene expression—is potentially the same as altering DNA itself. For those who have not followed recent trends in developmental and evolutionary biology this claim may seem startling, but it is now widely accepted that genotypic variation is only one of the sources of phenotypic change (Jablonka and Lamb 2005).[2]

11.2.2 The Technical Dimension

In order to understand the effects that IVF techniques can have on development, we need to be familiar with the various steps or phases of intervention that IVF typically entails. Prior to conception, the principal phases are pituitary suppression, controlled ovarian stimulation, monitoring of oocyte maturation, retrieval of oocytes, classification and culture of oocytes, and collection and preparation of semen. These are the steps by which gametes are collected, selected, and prepared. After the preparation of the gametes, the interaction between gametes can occur through IVF, sometimes involving a special technique called Intracytoplasmic Sperm Injection (ICSI). For ICSI, a single sperm is selected for direct injection into an egg, while in the case of conventional IVF many sperm are placed next to an egg in a dish.

[2] It is important to note that the importance of the new epigenetic model of development for discussions of IVF techniques does not depend on the highly controversial status of the embryo as a person. Regardless of when we believe personal status is achieved, the epigenetic model changes how we understand the genetic and developmental basis of the health of future children, and this is the relevant issue under discussion here.

After conception (whether by ICSI or conventional IVF), subsequent phases involve the control of embryo development and uterine transfer. For the control of embryo development, the In Vitro Culture System addresses metabolic needs of the pre-implantation embryo after the oocyte (egg) has been fertilized. As these needs change according to developmental stage—that is to say, virtually every 24 h—the culture medium in which the embryo is growing must be changed regularly. Currently, this phase involves the use of "sequential culture media," which attempt to imitate the growth factors supplied by the mother-to-be during the embryo's journey along the fallopian tube to the uterus.

From now on, when I refer in a general way to IVF, I will be referring to *all* phases of intervention related to the preparation of gametes, fertilization, and the control of embryo development, and I will indicate the use of ICSI where applicable. In addition to these techniques, an embryonic biopsy and a pre-implantation genetic diagnosis (PGD) may also be carried out. This last phase is optional, but, as stated above, I will also analyse its impact on development. In short, the full multiphase view of IVF encompasses the various techniques by which fertilization is induced, embryo development is monitored, and the embryos are analysed for selection and transferral to the uterus. For now, it can be said that the overall objective of the entire process is to obtain the maximum number of healthy embryos for transferral. After transferral, the goal is that at least one embryo will achieve implantation and lead to a full-term delivery.

11.3 The Impact of ART Techniques on Development

Recent advances in developmental biology, especially those related to the epigenetic dimension described above, are now enabling us to understand the impact that these techniques have on gametes (Gs) and embryos (As). Let us start with the effect of IVF techniques on Gs. As indicated in the previous section, the maturation of the gametes is a critical phase for attaining the proper conditions of genetic configuration and cellular composition that are, in turn, essential to successful fertilization and healthy development. Regardless of DNA sequence, without proper cellular composition a gamete is not disposed for the formation of a zygote or for the optimum development of the embryo. Moreover, it is important to realize that the IVF phases of ovarian stimulation and in vitro manipulation of gametes precisely coincide with this crucial epigenetic reprogramming stage of gamete maturation. Recall that during gametogenesis the DNA of the ovum and the spermatozoon acquire certain peculiarities at the epigenetic and cytoplasmic levels, peculiarities that affect gene configuration and thereby affect gene expression over the entire course of development. Recent studies suggest that, at this stage, techniques related to ovarian stimulation (Denomme and Mann 2012; Anckaert et al. 2013), the cryopreservation of the gametes (Kläver et al. 2012) and the oocyte's in vitro maturation (Monzo et al. 2012) can all have a serious, adverse impact on the genetic and/or epigenetic traits of the gametes that are necessary for the healthy development of the embryo, as well as altering the energy metabolism of the oocytes (Krisher 2004; Ertzeid and Storeng 2001).

With these issues in mind, let us now analyse how IVF techniques affect an embryo (A). It has been shown that the culture media used during embryonic control (pre-implantation) can affect the expression of imprinted genes (Dupond and Sifer 2012). This phase of intervention is characterized by constant chemical induction with a potentially adverse impact on global gene expression, which in turn can compromise embryo physiology, development and viability (Gardner et al. 2013). Moreover, studies have shown that nutritional manipulations during blastocyst formation, such as culturing the embryo in media, have been associated with anomalies at the cleavage stage (Alikani et al. 2000). In addition, in the case of ICSI, the technique by which a single sperm is selected and injected directly into an egg, epidemiological studies in animal models have established associations between the use of ICSI and diverse health problems, ranging from metabolic complications in the offspring to abortions, congenital malformations, and imprinting disorders (Sánchez-Calabuig et al. 2014). Other studies show that altered gene expression occurs in placental tissue of ICSI patients when compared with the placental tissue of spontaneous pregnancies in both non-imprinted and imprinted genes (Nelissen et al. 2014). ICSI was also associated with a higher level of methylation in an imprinted gene in comparison with standard IVF and spontaneous conception (Whitelaw et al. 2014). All these studies suggest that the stress inflicted on oocytes by the ICSI procedure may cause epigenetic defects in embryos (Tarín et al. 2014).

Now let us turn to possible damage and subsequent health risks involved in the optional phase of IVF mentioned previously, PGD (pre-implantation genetic diagnosis). To conduct a PGD, it is necessary to perform a biopsy. First, the zona pellucida that covers the embryo is perforated using a laser beam and then one or two cells are detached from the embryo by aspiration. These cells provide the genetic information necessary for selecting an embryo. We know that the embryo from which one or two cells have been removed recovers insofar as cell proliferation is seen to continue. However, the disruption in the zona pellucida provokes changes in the size of the embryo, as well as slows down development (Kirkegaard et al. 2012), affecting the differentiation of cells into tissues in the short, medium and long term. In fact, certain studies suggest that the biopsy procedure and, by extension, any other physical or chemical interference at this stage, however slight, poses the risk of altering the embryo and subsequent foetal development and, as a result, may have negative long-term effects in addition to its short-term impact (Tarín et al. 2014; Dupond and Sifer 2012). In fact, in studies of mice there is no doubt that interventions such as the removal of a portion of embryo mass and exposure to different solutions and manipulations before or during the biopsy process are associated with epigenetic alterations (Yu et al. 2009; Zhao et al. 2013; Wu et al. 2014). In summary, as stated by a recent review,

> "[t]he in vitro manipulations applied during IVF/ICSI treatments coincide with the developmental stages during which genome-wide epigenetic reprogramming occurs. Consequently, these manipulations may induce epigenetic disturbances in both male and female embryos, with serious consequences for pre- and post-implantation embryonic development" (Tarín et al. 2014).

11.4 Implications of ART for the Well-Being of the Offspring

In light of the effects that IVF techniques can have on Gs and A, let us now examine how such interventions can affect A*. There are several studies in animal models suggesting that IVF and related techniques can have an effect on later development (Young et al. 1998; Sinclair et al. 1999; Young et al. 2001; Rooke et al. 2007; Feuer et al. 2013). Moreover, the specific effects of IVF on the human embryo are now coming to light at the molecular level. As epigenetic alterations are increasingly recognized as causes of human disease (i.e. Gluckman et al. 2009; Dempster et al. 2011; Champagne 2012; Lister et al. 2013; Lee and Alisch 2012; Lorthongpanich et al. 2013), and some of these alterations are likely to occur during the periconceptional period when genomic imprints and the epigenome are most vulnerable (Padhee et al. 2015), concern for the epigenetic effects of interventions during this period seems warranted. Indeed, the available evidence supports the inference that IVF techniques have the potential to harm the health of the future child, this potential for harm must be taken into account.[3]

Within the scientific community, a growing concern for the effects of IVF techniques can be discerned in publications that focus on the risks posed to the development of the embryo and the future health of offspring (Dupond and Sifer 2012; Kohda and Ishino 2013; Hansen et al. 2013; Tararbit et al. 2013; Tarín et al. 2014; Padhee et al. 2015). It seems clear that IVF techniques constitute an intrusion into the developmental process, and can be expected to influence the epigenome of the resultant offspring (el Hajj and Haaf 2013), but what has science said up to now about the health of the offspring?

Scientific research has now accumulated enough cases for us to analyse the long-term health of these children. There are currently over a 100 published studies relevant to this issue, including eight reviews and four large, recent cohort studies that compare birth defects in ART and non-ART children. The following table (Table 11.1) brings together some of the research published about the relationship between IVF and ICSI and health, and shows the pathologies with an increased risk in ART children compared to those spontaneously conceived.

It is important to note that because large cohorts and metadata do not usually distinguish between offspring who were analysed by PGD and those who were not the consequences of this phase have yet to be fully determined. However, if ARTs that do not involve gamete manipulation (excluding IVF and ICSI as well as PGD), and thus can be considered as less invasive types of assisted reproduction, are still associated with an increased risk of birth defects (Davies et al. 2012), then it seems that PGD is not the main factor influencing the health problems of ART-produced offspring (Desmyttere et al. 2009).

[3] In fact, sensitivity is so high during the periconceptional period that deficiencies or excesses of a range of macro and micronutrients during the final stages of oocyte growth and post-fertilisation development can lead to impairments in foetal development and long-term offspring health (Sinclairand Watkins 2013).

Table 11.1 Pathologies with an increased risk in ARTs children compared to those spontaneously conceived

Year	Type	Pathologies with increased risk	Comparison	References
2002	Research	Cerebral palsy	ARTs vs NC	Stromberg et al. (2002)
2003	Research	Cancer-retinoblastoma	ART vs NC	Moll et al. (2003)
2004	Metadata	Major malformation	ART vs NC	Rimm et al. (2004)
2005	Metadata	Major malformation	ART vs NC	Hansen et al. (2005)
2005	Metadata	Pre-term birth and an increased rate of caesarean delivery	ART vs NC	McDonald et al. (2005a)
2005	Metadata	Pre-natal mortality	ART vs NC	McDonald et al. (2005b)
2005	Metadata	Major birth defects	ISCI vs IVF	Lie et al. (2005)
2006	Research	Cancer-hepatoblastoma	ART vs NC	McLaughlin et al. (2006)
2006	Research	Cerebral palsy	ARTs vs NC	Hvidtjorn et al. (2006)
2007	Research	Metabolic disease	IVF vs ISCI vs NC	Ceelen (2007)
2009	Research	Septal heart defects, cleft lip with or without cleft palate, esophageal atresia, hypospadias and norectal atresia	ARTs vs NC	Reefhuis et al. (2009)
2009	Research	Imprinting disorders	ARTs vs NC	Manipalvitatn et al. (2009)
2011	Metadata	Adverse outcomes and perinatal death	ARTs vs NC	Rossi and D'Addario (2011)
2012	Metadata	Obstetric and perinatal complications	ARTs vs NC	Pandey et al. (2012)
2012	Research	Chromosomal anomalies, cerebral palsy, general birth defects	ISCI vs IVF vs NC	Davies et al. (2012)
2012	Metadata	Birth defects	ARTs vs NC	Wen et al. (2012)
2012	Research	Cancer-leukemia	ARTs vs NC	Petridou et al. (2012)
2013	Metadata	Birth defects	ARTs vs NC	Hansen et al. (2013)
2013	Large cohorts	Cancer – rhabdomyosarcoma and hepatoblastoma	ARTs vs NC	Williams et al. (2013)
2014	Large cohorts	Cancer – central nervous system tumors and malignant epithelial neoplasm	ARTs vs NC	Sudnh et al. (2014)
2014	Large cohorts	Major congenital anomalies	ARTs vs NC	Pelkonen et al. (2014)
2014	Large cohorts	Imprinting disorders	ARTs vs NC	Lazaraviciute et al. (2014)
2014	Research	Epigenetic changes associated with health and fertility of future generations	ISCI vs IVF vs NC	Whitelaw et al. (2014)
2015	Metadata	Genitourinary malformation	ISCI vs IVF	Massaro et al. (2015)
2015	Review	Cardiovascular dysfunction	RTs vs NC	Scherrer et al. (2015)
2016	Large cohorts	Cancer – Leukemia and Hodgkin's lymphoma	ARTs vs NC	Reigstad et al. (2016)

Legend: *ARTs* ARTs children with no distinction between IVF and ISCI, *ISCI* Child born via ISCI, *IVF* Child born via IVF, *NC* Child naturally conceived

The current widespread use of ICSI, together with its likely increase, calls for an analysis of the possible health risks associated with this technique in comparison with standard IVF. A meta-analysis shows that the pooled estimate of the risk of a major birth defect showed a 1.12-fold increase after use of ICSI when compared with standard IVF (Lie et al. 2005). A more recent systematic review has concluded that ICSI is associated with a slightly higher risk of genitourinary malformation in offspring compared with standard IVF (Massaro et al. 2015), and another systematic review and meta-analysis shows that there is an increase in imprinting disorders in children conceived through ICSI as compared with children conceived spontaneously (Lazaraviciute et al. 2014). Although IVF is associated with reduced risk for all birth defects when compared with ICSI, both IVF and ICSI are associated with a significantly increased risk when compared with births to fertile women that did not involve assisted conception (Davies et al. 2012).

Finally, concerning how uterine transfer can affect the health of the offspring, recent studies show that transferring only one embryo per cycle during assisted reproduction would not only lower the rate of multiple pregnancies, but also have an important effect on the health of single babies (Henningsen et al. 2015).

It is also worthwhile to consider cardiovascular defects that result from ARTs, as these are currently the best-studied health outcomes. Studies show that epigenetic mechanisms are involved in the foetal programming of vascular dysfunction, and that IVF induces vascular dysfunction in the offspring (Valenzuela-Alcaraz et al. 2013; Rimoldi et al. 2014; Padhee et al. 2015). A recent review highlights emerging evidence pointing to adverse cardiovascular outcomes before and after birth in offspring conceived through IVF and related techniques in both human and animal studies (Padhee et al. 2015). Moreover, the article identifies the potential underlying causes as the hormonal and nutritional manipulation that follows superovulation and embryo culture and the lack of communication between tissues of the mother and the early embryo. Another investigation found that IVF, ICSI and ovulation induction are each associated with a 2.4-fold increased risk of tetralogy of fallot (a congenital heart defect), after adjusting for maternal age, occupation, geographic origin, paternal age and year of birth (Tararbit et al. 2013).

To conclude this review of the available evidence, the following quote from a recent review presents an overview of the current state of the question at hand—how ART techniques affect Gs, A and A*:

> Perturbations during oocyte maturation, such as those that may occur during parts of ART (ovarian hyperstimulation, in vitro maturation and in IVF) have been shown to reduce the quality of oocytes and embryo viability as well as alter energy metabolism of the oocytes. This has been shown to result in delayed embryonic development, increased abnormal blastocyst formation, foetal growth retardation, increased foetal loss, congenital malformations, imprinting disorders, and a range of postnatal growth and development disorders such as poor cognitive development, increased risk for neurological problems, cardiovascular diseases and respiratory tract infections (Padhee et al. 2015).

11.5 Discussion

11.5.1 Causes and Conjectures

Although research is still ongoing, the indication of accumulating evidence permits us to claim that children who were born through ARTs have a greater risk of suffering serious illnesses than those conceived naturally. The available data confirms that the concerns raised in the previous section are warranted: IVF is linked to potential harm in the short, medium and long term.

However, perhaps the problems that afflict parents who seek IVF—i.e. the conditions that underlie infertility—can also be associated with increased risk to children. This is a crucial point because, were such a connection to be established, IVF would no longer be seen as the cause (or main cause) of effects on the health of offspring, and instead the parents would be seen as the primary cause. Pointing to infertility as a significant cause of health problems in the child is therefore highly relevant to questions of parental responsibility, since it seems to remove responsibility from the clinics and assign it solely to the parents. But how plausible is this connection? A search turned up only one recent article that deals with the question in a systematic way (Tarín et al. 2014), and it provides only a commentary on existing data. In the most relevant section of this article, Tarín and colleagues make use of the data reported in the 2011 Clinic Summary Report of all Society for Assisted Reproductive Technologies (SART) Member Clinics[4] to compare the ratios of success among women who have resorted to IVF after diagnoses of various kinds of infertility. Differences between the ratios of success lead the authors to conclude that some causes of infertility may contribute more than others to the low percentages of success observed in ART cycles. If this were so, it would be reasonable to infer that causes of infertility in parents might also be causes of health problems in children born from IVF.

It should be noted that the 2011 ART Fertility Clinic Success Rates Report states that "a comparison of clinic success rates may not be meaningful because patient medical characteristics and treatment approaches vary from clinic to clinic" (CDC 2013). Without more data about the variability of the sample we cannot say for sure what lies behind these differences in success rates. To be precise, a measure of statistical significance requires both the mean and the variability of the data, but only the mean is available in this case. It follows that we cannot claim that 'the data' indicates that certain causes of infertility contribute more than others to the development of the embryo.

While such questions of statistical significance may seem overly subtle, it is important to give them careful consideration. As is often the case in complex matters, evidence of correlation does not necessarily yield evidence of causation. To give an example, a cohort study claims to have detected an increase in imprinting

---

[4]The data can be consulted at https://www.sartcorsonline.com/rptCSR_PublicMultYea-r.aspx?ClinicPKID=0

disorders in children conceived though IVF and ICSI (Lazaraviciute et al. 2014). The authors, however, indicate that the study design does not support conclusions about the cause of this increase, as they were unable to determine the aetiology of infertility and they could not compare the data with a group of subfertile women to account for the effect of infertility per se on congenital abnormalities. Indeed, in the editorial of a leading journal, conclusions about ART techniques as causal factors for serious developmental problems (Valenzuela-Alcaraz et al. 2013)—although supported by experimental evidence and consistent with current scientific knowledge—are discounted as speculative (Leeson and Baskaran 2013). I would like to stress that, at this point, the idea that parents' infertility problems are the cause of health problems in IVF-born children is mere conjecture and to imply otherwise obscures the need for further evidence. As concluded by a recent study (Society of Obstetricians and Gynaecologists of Canada et al. 2014): "Until sufficient research has clarified the independent roles of infertility and treatment for infertility, couples should be counselled about the risks associated with treatment."

Even when the limitations of these studies are kept in mind, and their results are interpreted with all due caution, there can be no doubt that all available reviews and cohort studies indicate that IVF and related techniques can cause serious alterations in the development of the gametes and of the embryo and, consequently, have the potential to produce significant increases in the risk of illness or defects.

11.5.2 The Risks Are Real, But the Responsibility Is Not

It is interesting to note that, in the realm of assisted human reproduction, research reverses the usual path taken in other biomedical disciplines. Instead of first conducting research with animals in order to discover secondary effects that might follow from a specific treatment that is being developed for humans, investigations with mice are now helping us to understand the damage caused by ART techniques that have already been used on human subjects. Recent studies have begun to highlight the importance of experiments that use mammalian models to test the implications of these techniques for humans (Sánchez-Calabuig et al. 2014). Given the currently available evidence, why is there so little debate? In my opinion, there are two motives.

First, society is heavily invested in IVF techniques because it is widely believed that people with fertility problems should nonetheless have the opportunity to have *their own*, i.e. *genetically related*, children. This is why fertility clinics are allowed to offer their services even if their techniques are associated with risks to the children that will be born. From a certain perspective one can claim that it is preferable to have a child with health problems than not to have any child at all.

Second, debate has generally failed to arise because, in many of the countries where ARTs are permitted, during the first developmental stages the organism possesses a different legal status from that of the foetus and the child and, as a result, manipulations of the embryo (including its destruction) are widely permitted.

At any rate, and as we will see in what follows, discussions of parental responsibility have given rise to new ethical considerations of this issue. As Dondorp and de Wert have shown in this volume, a subtle distinction has been made between acting in order to protect the interests of the foetus (for example, the forced hospitalization of a pregnant woman who uses cocaine) and acting in order to protect the health of the future child and, in turn, this distinction has important ethical repercussions. In the first case, the protection of the foetus during the period legally stipulated as "viable" might threaten the woman's right to terminate the pregnancy; in the second case, it does not. Hence, in order to avoid any conflicts of interest and to safeguard the woman's right to terminate the pregnancy, it is claimed that prenatal protection concerns the future of the child, not the foetus, and that it only makes sense to protect the foetus when the expectant mother has the intention of carrying the pregnancy to term. On the basis of this reasoning, would it make sense to hold that the embryo and the gametes from which it proceeds are worthy of protection because they pertain to the future of the child, provided that the couple or single reproducer intend to have a child?

Even if we grant that in most situations embryos (much less gametes) are not legally recognized as children and, moreover, the parents who opt for the use of IVF techniques tend not to think of embryos as children, the question remains: if these techniques have the potential to affect the health of future offspring, should parents be concerned about this possibility? As Hens notes in her chapter, scientific advances in our understanding of development, especially in the field of epigenetics, have raised new challenges for concepts of parental responsibility. Epigenetics indicates that the health of future children can be affected not only by the habits of pregnant women (e.g. smoking), but also by the habits or the environment of women and men who have not yet conceived or even not yet decided to ever become parents. This may seem like an absurdly broad notion of parental responsibility, but our current knowledge of health risks demands that we at least confront the issue. For example, we are now faced with the question of how to understand responsibility in light of the knowledge that a 16 year-old boy who smokes may—perhaps drastically—increase the chances of his future offspring developing asthma (Hens in this volume). Going even further, we are confronted with questions about the intergenerational dimension of epigenetics: for instance, the environmental exposure of a pregnant woman to harmful chemicals could affect not only the germ cells of the foetus she is carrying but also the health of the children of that foetus, i.e. her grandchildren (Whitelaw et al. 2014; Hens in this volume). These are new questions for ethical reflection, and whether or not they warrant the adjective "parental," they certainly involve responsibility. If we can ask questions about the future effects of the habits of adolescents who have not even thought about having children, and about the effects of the "natural environment" of a pregnant woman on her future grandchildren, it is reasonable that prospective parents and society in general should reflect on the effects of invasive procedures and of IVF on gametes and embryos that are intended to be selected as future children.

To clarify this point, consider the following story about Mary and Bob, two prospective parents who are hyper-vigilant about all matters related to parental responsibility. Let's suppose that Mary has been well informed since grade school that certain eating habits are crucial for the quality of her ova and therefore for the health of her future offspring. Her parents followed the guidelines for her proper diet during adolescence, and they kept an eye out for any illegal risky behaviour (smoking tobacco, drinking alcohol, etc.) In addition, Mary and her family changed residence during her adolescence because pollution had turned their city centre into a "dangerous zone" that presented a high probability of negative consequences not just for her children, but for her grandchildren as well. As an adult, Mary is assured that she has taken every precaution insofar as care for her gametes is involved: she has never smoked and has never drunk more alcohol than the recommended amount, etc. Mary meets Bob, and Mary and Bob decide they want to form a family. Mary becomes pregnant, and diligently follows all recommendations about healthy diet and living habits, and does her best to minimise exposure to any harmful environments. After 9 months, the child is born with heart problems. Driven by a strong sense of accountability, Mary and Bob seek to determine the causes of this disorder and identify the responsible parties. After careful investigation, they determine that Bob's parents did not adequately supervise his diet during a summer that he spent in Spain, and also that the dosages prescribed by Mary's doctor for sugar and amino acids during the pregnancy were inadequate, and they take legal action. Then, with the indemnification money in hand, they decide to enrol in the best available intensive reproductive health program, and follow a rigorous daily program for an entire year. Having finished the course, and after several unsuccessful attempts to get pregnant, they decide to use IVF. Mary and Bob are curious about how the clinic ensures that the interventions undergone by their gametes and the embryo prior to implantation will maintain the optimal conditions for healthy development—conditions about which both are extensively informed, and which they have always done their utmost to protect. They consider this matter to be no different from all the other possible risk factors which they have so diligently avoided. To their surprise, however, the clinic responds that, whatever the risks involved, they have no reason to concern themselves with any part of the IVF process, which lasts from the moment the mother-to-be undergoes ovarian stimulation until the foetus reaches 24 weeks. The clinic claims that because the gametes and the embryo are not yet persons, Mary and Bob are not yet parents, and there is nothing for which they bear any responsibility—and certainly nothing for which the clinic can later be held accountable. Imagine Mary and Bob's surprise, after all their efforts to protect their future children from unnecessary risks, and after strenuously upholding an extraordinarily broad and encompassing view of parental responsibility that extends even to diet and ambient air quality during adolescence, they are told that these stages of prefertilization preparation and early development, which are so intimately related to the formation of their child and so crucial to its short, medium, and long-term health, constitute a "no-man's land" of responsibility.

The motivation for this admittedly far-fetched scenario is to highlight the arbitrariness of drawing the boundaries of parental responsibility so that they exclude any possible consequences of IVF. It is striking—or, at least, it should be striking—that even when, as in the case of Mary and Bob, parental responsibility is stretched to include concern for the protection and welfare of the child before it's even born, *as determined by all possibly controllable circumstances*, this concern does not extend to the potentially damaging effects of ARTs on the health of future children, even though IVF is deliberately chosen and its risks are increasingly well established. Of course, the scenario of Mary and Bob is not realistic, as their sense of who is accountable for the health of their possible future offspring seems to extend the bounds of parental responsibility to include every possible influence on their reproductive health. Yet suppose we were to confine parental responsibility just to those who are presently, directly, and intentionally involved in the production of offspring: every scenario in which adults resort to ART involves risk factors that can be controlled as well as this present, direct, and intentional involvement. How could it not fall within the bounds of parental responsibility?

11.5.3 Consent to Risk and Damages: Medical and Parental Responsibility

The analysis performed in this article opens a sensitive topic which calls for further consideration: on the basis of the data available, it can be concluded that parents who resort to IVF and related techniques accept risks to their children and thereby implicitly consent to the damage that those techniques may cause. This seems to be a crucial issue for the responsibility of prospective parents. All parents must accept the risks that are connected with the development of their children. That is, whether or not they make use of IVF, parents have no choice but to assume certain risks associated with human development. However, in contrast with parents who conceive their children naturally, those who use IVF thereby expose their children to the potential damage that these techniques can cause and, consequently, must assume the additional risks that such manipulations entail. With respect to this issue, at the end of the last decade the scientific community was already cautioning that couples who are considering ART should be informed of its potential risks (Reefhuis et al. 2009).

In light of the evidence presented here, one wonders to what degree the parents who go to IVF clinics are appropriately informed about the potential damage to their children associated with each phase of the technique. Some might argue that parents who resort to ARTs have no choice but to accept the risks proceeding from IVF use. However, if they are well informed, parents *can* act so as to reduce or increase those risks. For instance, the mother-to-be can take folic acid before and during the first months of pregnancy in order to reduce the risks of neural tube defects. This is where the responsibility of prospective parents enters into the picture, and where it is undeniable that prospective parents should be rigorously well informed about the biological and technological aspects of embryo selection. It is only by knowing the impact of each phase of the technique on the gametes and embryos, and the resulting risks for health, that prospective parents are able to consent to the potential damages provoked, and can weigh the necessity of subjecting their gametes and future children to all of the technique's phases.

Here is another relevant example. Previously, I described how ovarian stimulation (Denomme and Mann 2012; Anckaert et al. 2013) and cryopreservation of the gamete (Kläver et al. 2012) can have an adverse impact on the genetic and/or epigenetic traits involved in establishing the molecular/cellular mechanisms for the optimum development of the embryo. If women/couples had this information and knew that these phases were not strictly necessary for a successful birth, perhaps they might think twice about consenting to them.

These questions situate us in a new terrain. Until now, ARTs have been largely evaluated within a framework for which the most important thing is to optimize costs relative to the success rate defined by a full-term delivery. The information under consideration introduces the question of how this optimization affects the long-term purpose of safeguarding or promoting the child's welfare, of looking out for the best interests of the child. In fact, this discussion has already begun. Recently, Anna-Karina Aaris Henningsen, based on her research on human reproduction, has suggested that a policy to transfer only one embryo per cycle during assisted reproduction would not only lower the rate of multiple pregnancies but also have an important effect on the health of individual offspring (Henningsen et al. 2015). This suggestion places the health of the offspring at the centre of the discussion: if it were to be followed, parental responsibility and medical responsibility would be better aligned, as both would seek to protect the health of the offspring (rather than maximize the number of full-term deliveries). Interestingly, in response to this statement, Antonio Requena, the medical director of the Valencia Infertility Institute (IVI), observed that the adoption of such policies is hindered by the fact that betting on a single embryo transfer will raise the cost of treatments (Diario ABC 2015). Here we see that considerations of cost-effectiveness are being placed on the same level as the health of the future offspring: would it be appropriate, in this case, to question medical responsibility?

11.6 Conclusion

Gametogenesis and the periconceptional period, defined as the period before and immediately after the time of conception, is a critical phase of early development (Louis et al. 2008; Padhee et al. 2015). IVF and related techniques are, by definition, invasive procedures that intervene in this critical phase: they involve physical and biochemical manipulation of the gametes and the fertilized cell that results from their interaction. ART techniques should therefore be considered as intrusions into the process of development which can influence the epigenome of the resultant offspring, and it is increasingly clear that such alterations can have long-lasting effects on the health of the individual.

Having presented and discussed the relevant evidence, we can now reformulate our initial goal in more precise terms. The aim of this paper has been to investigate whether IVF and related technics are indeed able to safeguard or promote the welfare of future children and whether they are capable of offering any future children the best chance for the best life, as required by the PPB. Let us analyse, then, whether the techniques employed with Gs and A do in fact safeguard and promote the future welfare of children created in this manner, and are capable of offering A* the best chance for the best life.

If we consider the consequences of standard IVF interventions—such as ovarian stimulation (Denomme and Mann 2012; Anckaert et al. 2013) in vitro maturation of the oocyte (Monzo et al. 2012), or the cryoconservation of the gametes—we find that it will be better for A* if the Gs which A comes from do not undergo these techniques, if what is desired is to safeguard A*'s health and offer her the best possible future. Moreover, if we consider the consequences of other interventions—such as transferring more than one embryo per cycle, ICSI, or the cryoconservation of the embryo—if we want the best possible A*, then prospective parents will have a moral obligation to avoid these interventions if the future health of the child(ren) is at stake. Finally, if embryo selection takes place through PGD, because of the considerable risk of harmful consequences, if we are looking out for the health of the future child, or seeking to give her the best possible life, we have a clear obligation to avoid this phase.

At the very least, we can conclude that prospective parents should be informed not only of the specific level of risk and potential damages associated with each IVF technique, but also of the fact that none of the above IVF techniques can be considered to be risk free. A responsible choice is an appropriately well-informed choice, and it does not seem that parents are currently provided with the information they need to make such choices about IVF. But it should also be said that, given the available evidence, there is reason to believe that none of these techniques can be reconciled with the responsibility of prospective parents to the promote welfare of future children and/or to offer them the best possible life.

References

Alikani, M., G. Calderon, G. Tomkin, J. Garrisi, M. Kokot, and J. Cohen. 2000. Cleavage anomalies in early human embryos and survival after prolonged culture in vitro. *Human Reproduction* 15: 2634–2643.

Anckaert, E., M. DeRycke, and J. Smitz. 2013. Culture of oocytes and risk of imprinting defects. *Human Reproduction Update* 19(1): 52–66.

CDC – Centers for Disease Control and Prevention. 2013. *2011 Assisted reproductive technology*, Fertility clinic success rates report. Resource document. Division of Reproductive Health. http://www.cdc.gov/art/ART2011/PDFs/ART_2011_Clinic_Report-Full.pdf. Accessed 24 Sept 2014.

Ceelen, M. 2007. Body composition in children and adolescents born after in vitro fertilization or spontaneous conception. *Journal of Clinical Endocrinology & Metabolism* 92: 3417–3423.

Champagne, F.A. 2012. Interplay between social experiences and the genome: Epigenetic consequences for behaviour. *Advances in Genetics* 77: 33–57.

Davies, M.J., V.M. Moore, K.J. Willson, P. Van Essen, K. Priest, H. Scott, E.A. Haan, and A. Chan. 2012. Reproductive technologies and the risk of birth defects. *New England Journal of Medicine* 366: 1803–1813.

Dempster, E.L., R. Pidsley, L.C. Schalkwyk, S. Owens, A. Georgiades, F. Kane, S. Kalidindi, M. Picchioni, E. Kravariti, T. Toulopoulou, R.M. Murray, and J. Mill. 2011. Disease-associated epigenetic changes in monozygotic twins discordant for schizophrenia and bipolar disorder. *Human Molecular Genetics* 20(24): 4786–4796.

Denomme, M.M., and M.R. Mann. 2012. Genomic imprints as a model for the analysis of epigenetic stability during assisted reproductive technologies. *Reproduction* 144(4): 393–409.

Desmyttere, S., M. Bonduelle, J. Nekkebroeck, M. Roelants, I. Liebaers, and J. De Schepper. 2009. Growth and health outcome of 102 2-year-old children conceived after preimplantation genetic diagnosis or screening. *Early Human Development* 85(12): 755–759.

Diario ABC – Noticias Salud. 2015. *Los niños 'probeta' no tienen más problemas de salud que los 'naturales'*. Resource Document. http://www.abc.es/salud/noticias/20150131/abci-seguridad-fertilizacion-vitro-201501301439.html. Accesed 2 Feb 2015.

Dupont, C., and C. Sifer. 2012. A review of outcome data concerning children born following assisted reproductive technologies. *ISRN Obstetrics and Gynecology* 2012: 405382.

el Hajj, N., and T. Haaf. 2013. Epigenetic disturbances in in vitro cultured gametes and embryos: Implications for human assisted reproduction. *Fertility and Sterility* 99(3): 632–641.

Ertzeid, G., and R. Storeng. 2001. The impact of ovarian stimulation on implantation and fetal development in mice. *Human Reproduction* 16: 221–225.

Feuer, S.K., L. Camarano, and P.F. Rinaudo. 2013. ART and health: Clinical outcomes and insights on molecular mechanisms from rodent studies. *Molecular Human Reproduction* 19(4): 189–204.

Gardner, D.K., R. Hamilton, B. McCallie, W.B. Schoolcraft, and M.G. Katz-Jaffe. 2013. Human and mouse embryonic development, metabolism and gene expression are altered by an ammonium gradient in vitro. *Reproduction* 146(1): 49–61.

Gluckman, P.D., M.A. Hanson, T. Buklijas, F.M. Low, and A.S. Beedle. 2009. Epigenetic mechanisms that underpin metabolic and cardiovascular diseases. *Nature Reviews Endocrinology* 5(7): 401–408.

Güell, F. 2013. *El estatuto biológico y ontológico del embrión humano: el paradigma epigenético del siglo XXI desde la teoría de la esencia de Xavier Zubiri*. Berna: Peter Lang.

Hansen, M., C. Bower, E. Milne, N. De Klerk, and J.J. Kurinczuk. 2005. Assisted reproductive technologies and the risk of birth defects – A systematic review. *Human Reproduction* 20(2): 328–338.

Hansen, M., J.J. Kurinczuk, E. Milne, N. De Klerk, and C. Bower. 2013. Assisted reproductive technology and birth defects: A systematic review and meta-analysis. *Human Reproduction Update* 19(4): 330–353.

Henningsen, A.A., M. Gissler, R. Skjaerven, C. Bergh, A. Tiitinen, L.B. Romundstad, U.B. Wennerholm, O. Lidegaard, A. Nyboe Andersen, J.L. Forman, and A. Pinborg. 2015. Trends in perinatal health after assisted reproduction: A nordic study from the CoNARTaS group. *Human Reproduction* 30: 710–716. doi:10.1093/humrep/deu345.

Hvidtjorn, D., J. Grove, D.E. Schendel, M. Vaeth, E. Ernst, L.F. Nielsen, and P. Thorsen. 2006. Cerebral palsy among children born after in vitro fertilization: The role of preterm delivery – A population-based, cohort study. *Pediatrics* 118: 475–482.

Jablonka, E., and M. Lamb. 2005. *Evolution in four dimensions – Genetic, epigenetic, behavioral, and symbolic variation in the history of life*. Cambridge, MA: MIT Press.

Kirkegaard, K., J.J. Hindkjaer, and H.J. Ingerslev. 2012. Human embryonic development after blastomere removal: A time-lapse analysis. *Human Reproduction* 27(1): 97–105.

Kläver, R., A. Bleiziffer, K. Redmann, C. Mallidis, S. Kliesch, and J. Gromoll. 2012. Routine cryopreservation of spermatozoa is safe-evidence from the DNA methylation pattern of nine spermatozoa genes. *Journal of Assisted Reproduction and Genetics* 29(9): 943–950.

Kohda, T., and F. Ishino. 2013. Embryo manipulation via assisted reproductive technology and epigenetic asymmetry in mammalian early development. *Philosophical Transaction Royal Society of London B* 368: 20120353.

Krisher, R.L. 2004. The effect of oocyte quality on development. *Journal of Animal Science* 82: E14–E23.

Lazaraviciute, G., M. Kauser, S. Bhattacharya, P. Haggarty, and S. Bhattacharya. 2014. A systematic review and meta-analysis of DNA methylation levels and imprinting disorders in children conceived by IVF/ICSI compared with children conceived spontaneously. *Human Reproduction Update* 20(6): 840–852.

Lee, E.R., and R.S. Alisch. 2012. Early-life disruption of epigenetic marks may contribute to the origins of mental illness. *Epigenomics* 4(4): 355–357.

Leeson, P., and T. Baskaran. 2013. "Assisted" reshaping of the fetal heart? *Circulation* 128: 1398–1399.

Lie, R.T., A. Lyngstadass, K.H. Ørstavik, L.S. Bakketeig, G. Jacobsen, and T. Tanbo. 2005. Birth defects in children conceived by ICSI compared with children conceived by other IVF-methods; a meta-analysis. *International Journal of Epidemiology* 34(3): 696–701.

Lister, R., E.A. Mukamel, J.R. Nery, M. Urich, C.A. Puddifoot, N.D. Johnson, J. Lucero, Y. Huang, A.J. Dwork, M.D. Schultz, M. Yu, J. Tonti-Filippini, H. Heyn, S. Hu, J.C. Wu, A. Rao, M. Esteller, C. He, F.G. Haghighi, T.J. Sejnowski, M.M. Behrens, and J.R. Ecker. 2013. Global epigenomic reconfiguration during mammalian brain development. *Science* 341: 6146.

Lorthongpanich, C., L.F. Cheow, S. Balu, S.R. Quake, B.B. Knowles, W.F. Burkholder, D. Solter, and D.M. Messerschmidt. 2013. Single-cell DNA-methylation analysis reveals epigenetic chimerism in preimplantation embryos. *Science* 341: 1110–1112.

Louis, G.M.B., M.A. Cooney, C.D. Lynch, and A. Handal. 2008. Periconception window: Advising the pregnancy-planning couple. *Fertility and Sterility* 89: e119–e121.

Manipalviratn, S., A. De Cherney, and J. Segars. 2009. Imprinting disorders and assisted reproductive technology. *Fertility and Sterility* 91: 305–315.

Massaro, P.A., D.L. MacLellan, P.A. Anderson, and R.L. Romao. 2015. Does intracytoplasmic sperm injection pose an increased risk of genitourinary congenital malformations in offspring compared to in vitro fertilization? A systematic review and meta-analysis. *The Journal of Urology* 193(5): 1837–1842.

McDonald, S.D., K. Murphy, J. Beyene, and A. Ohlsson. 2005a. Perinatel outcomes of singleton pregnancies achieved by in vitro fertilization: A systematic review and meta–analysis. *Journal d'obstétrique et gynécologie du Canada* 27(5): 449–459.

McDonald, S., K. Murphy, J. Beyene, and A. Ohlsson. 2005b. Perinatal outcomes of in vitro fertilization twins: a systematic review and meta–analyses. *American Journal of Obstetrics and Gynecology* 193(1): 141–152.

McLaughlin, C.C., M.S. Baptiste, M.J. Schymura, P.C. Nasca, and M.S. Zdeb. 2006. Maternal and infant birth characteristics and hepatoblastoma. *American Journal of Epidemiology* 163: 818–828.

Medical Protection Society. 2012. *Parental responsibility*. Resource document. http://www.medicalprotection.org/mps–factsheet–parental–responsibility.pdf. Accessed 20 Oct 2014.

Moll, A., S. Imhof, J. Cruysberg, A.Y. Schouten–van Meeteren, M. Boers, and F. Van Leeuwen. 2003. Incidence of retinoblastoma in children born after in-vitro fertilization. *Lancet* 361: 309–310.

Monzo, C., D. Haouzi, K. Roman, S. Assou, H. Dechaud, and S. Hamamah. 2012. Slow freezing and vitrification differentially modify the gene expression profile of human metaphase II oocytes. *Human Reproductiom* 27(7): 2160–2168.

Nelissen, E.C.M., J.C.M. Dumoulin, F. Busato, L. Ponger, L.M. Eijssen, J.L.H. Evers, J. Tost, and A.P.A. van Montfoort. 2014. Altered gene expression in human placentas after IVF/ICSI. *Human Reproduction* 29(12): 2821–2831.

Olsen, A. 2015. Punishing parents: Child removal in the context of drug use. *Drug and Alcohol Review* 34(1): 27–30.

Padhee, M., S. Zhang, S. Lie, K.C. Wang, K. Botting, I.C. McMillen, S.M. MacLaughlin, and J.L. Morrison. 2015. The periconceptional environment and cardiovascular disease: Does in vitro embryo culture and transfer influence cardiovascular development and health. *Nutrients* 7(3): 1378–1425.

Pandey, S., A. Shetty, M. Hamilton, S. Bhattacharya, and A. Maheshwari. 2012. Obstetric and perinatal outcomes in singleton pregnancies resulting from IVF/ICSI: A systematic review and meta-analysis. *Human Reproduction Update* 18(5): 485–503.

Pelkonen, S., A.-L. Hartikainen, A. Ritvanen, R. Koivunen, H. Martikainen, M. Gissler, and A. Tiitinen. 2014. Major congenital anomalies in children born after frozen embryo transfer: A cohort study 1995–2006. *Human Reproduction* 29(7): 1552–1557.

Petridou, E.T., T.N. Sergentanis, P. Panagopoulou, M. Moschovi, S. Polychronopoulou, M. Baka, A. Pourtsidis, F. Athanassiadou, M. Kalmanti, V. Sidi, et al. 2012. In vitro fertilization and risk of childhood leukemia in Greece and Sweden. *Pediatric Blood & Cancer* 58: 930–936.

Reefhuis, J., M.A. Honein, L.A. Schieve, A. Correa, C.A. Hobbs, and S.A. Rasmussen. 2009. The national birth defects prevention study assisted reproductive technology and major structural birth defects in the United States. *Human Reproduction* 24: 360–366.

Reigstad, M.M., I.K. Larsen, T.O. Myklebust, T.E. Robsahm, N.B. Oldereid, L.A. Brinton, and R. Storeng. 2016. Risk of cancer in children conceived by assisted reproductive technology. *Pediatrics* 137(3), e20152061.

Rimm, A.A., A.C. Katayama, M. Diaz, and K.P. Katayama. 2004. A meta-analysis of controlled studies comparing major malformation rates in IVF and ICSI infants with naturally conceived children. *Journal of Assisted Reproduction and Genetics* 21(12): 437–443.

Rimoldi, S., C. Sartori, E. Rexhaj, D. Cerny, R. von Arx, R. Soria, M. Germond, Y. Allemann, and U. Scherrer. 2014. Vascular dysfunction in children conceived by assisted reproductive technologies: Underlying mechanisms and future implications. *Swiss Medical Weekly* 144: w13973.

Rooke, J.A., T.G. McEvoy, C.J. Ashworth, J.J. Robinson, I. Wilmut, L.E. Young, and K.D. Sinclair. 2007. Ovine fetal development is more sensitive to perturbation by the presence of serum in embryo culture before rather than after compaction. *Theriogenology* 67: 639–647.

Rossi, A.C., and V. D'Addario. 2011. Neonatal outcomes of assisted and naturally conceived twins: Systematic review and meta–analysis. *Journal of Perinatal Medicine* 39(5): 489–493.

Sánchez-Calabuig, M.J., A.P. López-Cardona, R. Fernández-González, P. Ramos-Ibeas, N. Fonseca, R. Laguna, E. Pericuesta, A.A. Gutiérrez, and P. Bermejo-Álvarez. 2014. Potential health risks associated to ICSI: Insights from animal models and strategies for a safe procedure. *Frontiers in Public Health* 2: 241.

Savulescu, J. 2001. Procreative beneficence: Why we should select the best children. *Bioethics* 15: 413–426.

Savulescu, J., and G. Kahane. 2009. The moral obligation to create children with the best chance of the best life. *Bioethics* 23(5): 274–290.

Scherrer, U., E. Rexhaj, Y. Allemann, C. Sartori, and S.F. Rimoldi. 2015. Cardiovascular dysfunction in children conceived by assisted reproductive technologies. *European Heart Journal* 36: 1583–1589.

Sinclair, K.D., T.G. McEvoy, E.K. Maxfield, C.A. Maltin, L.E. Young, I. Wilmut, P.J. Broadbent, and J.J. Robinson. 1999. Aberrant fetal growth and development after in vitro culture of sheep zygotes. *Journal of Reproduction and Fertility* 116: 177–186.

Sinclair, K.D., and A.J. Watkins. 2013. Parental diet, pregnancy outcomes and offspring health: Metabolic determinants in developing oocytes and embryos. *Reproduction, Fertility and Development* 26(1): 99–114.

Society of Obstetricians and Gynaecologists of Canada, N. Okun, and S. Sierra. 2014. Pregnancy outcomes after assisted human reproduction. *Journal of Obstetrics and Gynaecology Canada* 36(1): 64–83.

Stromberg, B., G. Dahlquist, A. Ericson, O. Finnstrom, M. Koster, and K. Stjernqvist. 2002. Neurological sequelae in children born after in-vitro fertilisation: A population-based study. *Lancet* 359: 461–465.

Sundh, K.J., A.A. Henningsen, K. Källen, C. Bergh, L.B. Romundstad, M. Gissler, A. Pinborg, R. Skjaerven, A. Tiitinen, D. Vassard, B. Lannering, and U.B. Wennerholm. 2014. Cancer in children and young adults born after assisted reproductive technology: A Nordic cohort study from the Committee of Nordic ART and Safety (CoNARTaS). *Human Reproduction* 29(9): 2050–2057.

Tararbit, K., N. Lelong, A.-C. Thieulin, L. Houyel, D. Bonnet, F. Goffinet, B. Khoshnood, and EPICARD Study Group. 2013. The risk for four specific congenital heart defects associated with assisted reproductive techniques: A population-based evaluation. *Human Reproduction* 28: 367–374.

Tarín, J.J., M.A. García-Pérez, and A. Cano. 2014. Assisted reproductive technology results: Why are live-birth percentages so low? *Molecular Reproduction & Development* 81: 568–583.

Valenzuela-Alcaraz, B., F. Crispi, B. Bijnens, M. Cruz-Lemini, M. Creus, M. Sitges, J. Bartrons, S. Civico, J. Balasch, and E. Gratacós. 2013. Assisted reproductive technologies are associated with cardiovascular remodeling in utero that persists postnatally. *Circulation* 128: 1442–1450.

Wen, J., J. Jiang, C. Ding, J. Dai, Y. Liu, Y. Xia, J. Liu, and Z. Hu. 2012. Birth defects in children conceived by in vitro fertilization and intracytoplasmic sperm injection: A meta-analysis. *Fertility and Sterility* 97(6): 1331–1337.

Whitelaw, N., S. Bhattacharya, G. Hoad, G.W. Horgan, H. Hamilton, and P. Haggarty. 2014. Epigenetic status in the offspring of spontaneous and assisted conception. *Human Reproduction* 29(7): 1452–1458.

Williams, C.L., K.J. Bunch, C.A. Stiller, M.F.G. Murphy, B.J.W. Botting, H. Wallace, M. Davies, and A.G. Sutcliffe. 2013. Cancer risk among children born after assisted conception. *The New England Journal of Medicine* 369: 1819–1827.

Wu, Y., Z. Lv, Y. Yang, G. Dong, Y. Yu, Y. Cui, M. Tong, L. Wang, Z. Zhou, H. Zhu, Q. Zhou, and J. Sha. 2014. Blastomere biopsy influences epigenetic reprogramming during early embryo development, which impacts neural development and function in resulting mice. *Cellular and Molecular Life Sciences* 71: 1761–1774.

Young, L.E., K.D. Sinclair, and I. Wilmut. 1998. Large offspring syndrome in cattle and sheep. *Reviews of Reproduction* 3: 155–163.

Young, L.E., K. Fernandes, T.G. McEvoy, S.C. Butterwith, C.G. Gutierrez, C. Carolan, P.J. Broadbent, J.J. Robinson, I. Wilmut, and K.D. Sinclair. 2001. Epigenetic change in IGF2R is associated with fetal overgrowth after sheep embryo culture. *Nature Genetics* 27: 153–154.

Yu, Y., J. Wu, Y. Fan, Z. Lv, X. Guo, C. Zhao, R. Zhou, Z. Zhang, F. Wang, M. Xiao, L. Chen, H. Zhu, W. Chen, M. Lin, J. Liu, Z. Zhou, L. Wang, R. Huo, Q. Zhou, and J. Sha. 2009. Evaluation of blastomere biopsy using a mouse model indicates the potential high risk of neurodegenerative disorders in the offspring. *Molecular and Cellular Proteomics* 8(7): 1490–1500.

Zhao, H.C., Y. Zhao, M. Li, J. Yan, L. Li, R. Li, P. Liu, Y. Yu, and J. Qiao. 2013. Aberrant epigenetic modification in murine brain tissues of offspring from preimplantation genetic diagnosis blastomere biopsies. *Biology of Reproduction* 89(5): 117.

Chapter 12
Should Parents Take Active Steps to Preserve Their Children's Fertility?

Daniela Cutas

Abstract It has been argued that, when there is a probable imminent risk of loss of children's fertility, their parents should take active steps to preserve their reproductive potential if possible – or even that children have a right to such interventions being undertaken on them on their behalf, as an expression of their right to an open future. In this chapter, I explore these proposals and some of their implications. I place the discussion of fertility preservation for children into the more general context of the choices that parents might have to help keep their children's future reproductive (and parenting) choices open. I discuss the role of desert and fairness in arguments for fertility preservation and their relevance for framings of infertility in general, as well as the relation between having a (slight) possibility to reproduce and becoming a parent.

Keywords Fertility preservation • Infertility • Children • Fairness • Desert • Cancer

12.1 Introduction[1]

The ever increasing capability developed in the last few decades to preserve human reproductive material outside the body has created a host of ethical dilemmas: who if anyone *owns* such material? Who can use it and who ought to decide how it is used or disposed of? What ought to happen to it if the person from whom it was collected has died? Where children are concerned, these developments take place at

[1]Work on this chapter was supported by the Swedish Research Council (grant number 421-2013-1306).

D. Cutas (✉)
Department of Historical, Philosophical and Religious Studies, Umeå University, SE-901 87 Umeå, Sweden

Department of Philosophy, Linguistics and Theory of Science, University of Gothenburg, Gothenburg, Sweden
e-mail: daniela.cutas@umu.se

K. Hens et al. (eds.), *Parental Responsibility in the Context of Neuroscience and Genetics*, International Library of Ethics, Law, and the New Medicine 69,
DOI 10.1007/978-3-319-42834-5_12

a time in which more children survive conditions, such as cancer, which only a few years or decades ago they would not have survived. Unfortunately, their capacity to reproduce is often compromised in this process. These two arms of medical progress raise questions as to which sort of, if any, reproductive decisions ought to be taken on children's behalf when their fertility is threatened. If certain conditions or life-saving treatments risk rendering the child infertile by the time she reaches adulthood, should parents, or someone else, take measures to preserve her reproductive potential?

For children too young to be able to give informed consent, their parents (or legal guardians) are *prima facie* expected to make decisions on their behalf. The legitimacy of parental choices for their children's care is dependent on the condition that these choices are taken in the children's interest, and are likely to benefit the children.[2] As we will see further in this chapter, this assumption is in line with the literature discussing fertility preservation for children: no reasons other than the interests of children have been advanced as primary motivating arguments either for or against measures towards fertility preservation for children – though, as we will see further in this chapter, some such underlying reasons (such as the parents' own desire to become grandparents) have been suspected.

Depending on the age and the sex of the children as well as the time available, fertility preservation may include procedures that can range from minimally invasive, such as collection of sperm from adolescent boys, to invasive ones such as collection of ovarian or testicular tissue. The reproductive material can either be transplanted back onto the child at a later date (in the case of ovarian or testicular tissue), or stored until the child has reached adulthood and solicits access to it to use via e.g. artificial insemination or assisted reproductive technologies (ARTs) (in the case of sperm).

The option of fertility preservation for children has mostly been discussed in the context of cancer treatment. Although chances of recovery for children suffering from cancer have improved considerably over the last decades, chemotherapy and radiation treatment still often leave the patient infertile – even though efforts are made to avoid this effect. These children will grow up to be adults who may develop a wish to reproduce but find themselves unable to do so. That there may be possibilities to preserve their reproductive potential has raised the issue of whether parents should explore these possibilities and take active steps towards the preservation of this capacity in their children.

I have discussed fertility preservation for children elsewhere (Cutas and Hens 2015). In this chapter, I briefly present the current context in which fertility preservation for children is discussed, as well as the arguments presented in favour of the claim that parents should take fertility preservation measures for their children. I present the arguments from the children's right to an open future, as it has been invoked in this context, and some of the shortcomings of current proposals. I explore other implications that such a right might have, such as that children's fertility

[2] There may be disagreement on whether this condition really lies at the core of the legitimacy of parental decision making. For the purposes of this chapter, I will assume that it does.

should be preserved *even* or *especially* in absence of immediate threats, or even if the child objects. I discuss the framing of the discussion around fertility preservation in the context of cancer in terms of fairness and *clinical* or *undeserved* infertility, and highlight some of the pitfalls of this framing. Throughout the chapter, unless otherwise specified, I use "fertility" loosely to denote capacity to reproduce, and "infertility" the lack of this capacity – as we will see later in the chapter, defining infertility is no easy task (for more on this, see Smajdor and Cutas 2015) and can depend on what purpose one needs the definition for.

12.2 The Context of Arguments for and Against Fertility Preservation for Children

Most of the literature discussing fertility preservation for children focuses on children facing potentially fertility impairing cancer treatment, chemotherapy and radiation treatment in particular. These treatments are also used in the context of other conditions, such as sickle-cell anaemia (Demeestere et al. 2015; Imbert et al. 2014). Yet another context in which the issue of fertility preservation for children has been raised is that of conditions such as Klinefelter syndrome, which is characterized by male individuals having one or more extra X chromosomes, and which renders these persons infertile in almost all cases by the time they reach adulthood. Should the syndrome be identified early enough, the decision to remove and preserve reproductive material from the child can be made and implemented (Herlihy et al. 2011; Mehta and Paduch 2012; Gies et al. 2012; Cutas and Hens 2015).

As children survive cancer, they become adults who may develop a wish to become parents. However, they are often left infertile subsequent to cancer treatment, even as efforts are made to prevent it (Thomson et al. 2002). In some cases, it may be because their gonads have had to be removed during treatment, and in some because their germ cells have been damaged during chemotherapy and radiation treatment (Thomson et al. 2002; Simon et al. 2005; Fallat and Hutter 2008). Depending on their age before the start of the treatment, measures can and in some cases have been taken to preserve these children's fertility. These measures can range from long established sperm preservation techniques to much more invasive procedures, such as ovarian tissue cryopreservation. The latter has been performed on girls as young as 2 years old (Rochman 2011; Quinn et al. 2012). The first successful case of fertility restoration after ovarian tissue removed during childhood and cryopreserved has recently been reported (Demeestere et al. 2015): in this case, the ovarian tissue of a patient with sickle-cell anaemia was removed when she was 14 years old, and was transplanted back when she was 25. She gave birth 2 years later.

Cancer survivors can suffer from significant distress because of their infertility (Rieker et al 1990; Schover 2005), and some of them place even more value on genetic reproduction than the general population, or put the pain of infertility almost on a par with the pain of confronting cancer (Schover 2005). When they have lost their fertility without having been informed about this risk and without having been offered fertility preservation, they feel wronged (Schover 2005). Against this background, discussions have emerged regarding the parents' entitlement, or even moral duty, to take measures to preserve their children's fertility (Grundy et al. 2001a, b; Schover 2005; Jadoul et al. 2010) – with some commentators claiming that children may have a right that such measures are taken on their behalf (Jadoul et al. 2010; Larcher 2012).

Among men who have had reproductive material stored before cancer treatment, some studies have found that only very few have used the material thus safeguarded, and of those, even fewer have produced offspring: the usage has been at around 6–9 % of survivors, with about half of these cases resulting in live births (Kelleher et al. 2001; Magelssen et al. 2005; van Casteren et al. 2008; Bizet et al. 2012). This very low usage rate, from cases in which the survivors themselves chose to have reproductive material preserved, suggests that the percentage of future use by the children of whom reproductive material has been stored might not be very high. In some of these cases, the survivors may have decided not to reproduce after all, and in some they may not have needed the material stored because they were fertile despite cancer treatment (Saito et al. 2005). Moreover, the psychological comfort that fertility preservation may provide adults who wish to become parents is arguably less salient in the case of children (Cutas and Hens 2015). However, this does not invalidate the case for fertility preservation in general, or for children in particular. That they may or may not eventually use the material stored is their choice – and that is what fertility preservation seeks to preserve: the possibility to choose whether or not to reproduce. Conversely, not preserving their fertility makes it much more likely that they will not have this choice.

Fertility preservation for children raises a number of specific medical as well as ethical problems. Many of the techniques being proposed or even undertaken on children are highly experimental. Ovarian tissue transplants have led to few successful births (Dittrich et al. 2012; Demeestere et al. 2015), and testicular tissue is yet to lead to any successful births (Poirot and Schubert 2011). Procedures such as these are invasive: any invasive surgical procedure has risks, and may add another layer of strain onto the child, at a moment when her life is in danger. The distress of cancer diagnostics and of the prospect of cancer treatment are the background against which these discussions about and measures towards fertility preservation take place. Moreover, it seems that there is a possibility that cancer cells will be reintroduced in the recipient's body together with the reproductive material in the case of techniques that presuppose subsequent transplantation back onto the patient (Grundy et al. 2001a, b; Fujita et al. 2005; Jahnukainen et al. 2006). In the absence of consent from children too young to give it, any risk to the child's wellbeing for the sake of a chance of genetic reproduction has to be considered against the likely benefits, in the difficult decision-making process of whether to proceed with fertility preservation or not.

Clearly, *which* methods are used and in which circumstances is very important. Whereas, for example, sperm collection from a teenage boy with his assent is a non-invasive method with a relatively high likelihood of adequate preservation of the material and success if subsequently used in reproduction, invasive methods such as ovarian or sperm tissue cryopreservation from very young children are much more problematic. In practice, this will mean that whether parents *should* undertake fertility preservation measures will depend on the particulars of each case (Cutas and Hens 2015).

Lastly, when parents are in a situation in which they can act to preserve their children's fertility, there is a distinct possibility that in doing so, their concern for their children's reproductive future is mixed with concern for their own grandparenting future. In opting for fertility preservation for their children, they might (a) project their own preferences (to reproduce) onto their children, and (b) try to secure a chance to become genetic grandparents. The latter possibility has been raised by several researchers (Nisker et al. 2006; Dolin et al. 2010; Bourke et al. 2014). The worry here is that parental decisions in these matters are not necessarily motivated by and reflective (only) of children's interests (Nisker et al. 2006) – and therefore, not always entirely legitimate.

12.3 Children's Right to Fertility Preservation

It has been claimed that measures should be taken to remove and store reproductive material from children for future use (if and when they choose to use it), and that they have a right to such procedures being undertaken: their right to an open future (Grundy et al. 2001a, b; Schover 2005; Jadoul et al. 2010; Larcher 2012). Not undertaking such measures deprives them of the choice to reproduce genetically. In the words of Jadoul et al.

> a child's right to fertility preservation is now acknowledged in the bioethical literature as a "right in trust", to be safeguarded until the child reaches adulthood. Unfortunately, such rights are sometimes violated in advance, before the child is in a position to exercise them (Feinberg 1992; Davis 1997). If the medical risk is acceptable, and the resources are available, it is an individual's ethical right to request fertility preservation. On the other hand, it is our ethical obligation to act "in the child's best interest". In order to determine the child's best interest, it is important to analyze the efficacy, feasibility and risks of the technique applied, particularly as this technology may still be seen as experimental and unproven. (Jadoul et al. 2010: 621)

It has been further claimed that doing nothing and allowing that children suffer from avoidable infertility subsequent to having suffered from cancer is "doubly unjust": not only have these children struggled with a life-threatening disease and have been subjected to difficult treatments, but they have been left infertile as well (Grundy et al. 2001b). Furthermore, some commentators have said that

> [i]t seems intuitively unfair that those rendered infertile by necessary treatment should bear the additional burden of cost for ART, especially if the latter is ethically justifiable. (Larcher 2012: 2)

Larcher defines infertility as "the inability to conceive after 1 year or more of unprotected intercourse", and sterility as the "inability to conceive without clinical intervention" (Larcher 2012: 1). These definitions are interesting because they both miss what most fertility preservation techniques are likely to offer: oocyte cryopreservation, for example, might offer a chance of genetic reproduction, but would not correct either sterility (because conception with the rescued oocytes requires clinical intervention) or infertility (because conception would not result from unprotected intercourse but, again, from clinical intervention). Only measures that would allow the return of the capacity to reproduce without clinical intervention, and/or via sexual intercourse (such as reproductive tissue transplants: though these *represent* clinical intervention prior to developing the ability to conceive!) could be described as either avoiding sterility or restoring fertility assuming Larcher's definitions.

Furthermore, according to Larcher,

> [t]he provision of ART in these circumstances is grounded in the moral obligation to respect the autonomous choices of individuals, which is fundamental to a free society. In general we do not prevent adults from exercising freely made reproductive choices unless to do so poses serious risks of harm to the children conceived or to others (Larcher 2012: 1).

Larcher starts from these considerations in his argument in favour of the "ethical obligation to preserve fertility in the face of all therapies that might adversely affect it" (the title of Larcher's 2012 article). This starting point is interesting because it collapses a negative right to reproduce into a positive one: that is, while Larcher presents this right as that of not being prevented from exercising freely made reproductive choices, he uses it to ground a right to fertility preservation. The right to fertility preservation, however, is not the same as not being prevented from making choices: instead, it is being proposed as grounding an ethical obligation onto others to take active measures to maintain certain reproductive choices. Classic examples of the negative right to reproductive choice are access to contraception and not being interfered with in one's reproductive projects (e.g. not being denied access to ARTs, not being subjected to forced sterilisation). Larcher's case for the ethical obligation to preserve fertility, however, is an example of the positive interpretation of the right to reproductive choice: a right that requires intervention from someone else.

When children's "right to an open future" is used to ground the right to fertility preservation (such as in the text above by Jadoul et al.), reference is often made to Joel Feinberg's formulation (1980, 1992): the original use of the term. However, the right to an open future in its original formulation does not straightforwardly support a right to fertility preservation. In Feinberg's words, the right to an open future is an anticipatory autonomy right that

> can be violated "in advance", so to speak, before the child is even in a position to exercise them. The violating conduct guarantees *now* that when the child is an autonomous adult, certain key options will already be closed to him. His right while he is still a child is to have these future options kept open until he is a fully formed, self-determining adult capable of deciding among them (Feinberg 1992: 77).

So far, it looks like Feinberg's (and Davis' subsequent development) concept fits the case for fertility preservation. However, the right that Feinberg described was negative in nature, rather than positive, and concerned mostly what parents ought *not to do* to their children, rather than what they should do to them for the sake of their own future. Feinberg's example was that of parents' choice to withdraw their children from school, on religious grounds. Davis' examples were of parents deliberately seeking to have children with certain disabilities (such as deafness) and, again, of parents wishing to restrict their children's futures on religious grounds; she sees these as examples of "a conflict between parental autonomy and the child's future autonomy" (Davis 1997: 8). Thus, when advancing the right to an open future, both Feinberg and Davis were concerned with *limiting* parental action, rather than calling for it. Feinberg's and Davis' arguments seem to apply well to ground a duty to try to preserve children's fertility while providing the treatment (for example by shielding the reproductive organs from harm, if that is possible), but do not straightforwardly extend to invasive measures to remove and preserve reproductive material. As in the case of the right to reproductive choice above, the right to an open future is, in the context of fertility preservation for children, given a *different*, positive interpretation. It is plausible that a good case can be made for the right to an open future grounding fertility preservation for children: however, that case is not made by invoking Feinberg and Davis' arguments.

Lastly, if children have a right to fertility preservation as an expression of their right to an open future, then perhaps it should not be up to their parents whether or not this right should be exercised[3]: instead, it should be protected by the state. The legitimacy of parental authority is questioned when their decisions threaten the exercise of their children's rights. If fertility preservation is a children's right, then parents should not be allowed to waive it on behalf of their children. Such a conclusion is far-reaching: should health care professionals intervene and undertake fertility preservation procedures on children? Should there be age limits, or limits dependent on the likely outcome (i.e. does the right require low risk, better success rates procedures, rather than experimental ones)? What risks are acceptable in the process of exercising children's right to an open future on their behalf? How should the cases in which parents are opposed to the procedures be handled? All these are important questions that need to be discussed whether or not children have a right to fertility preservation. If they do, procedures for ensuring the exercise of this right on their behalf should be in place. If they don't, their interests should likewise be safeguarded – if needed, by safeguarding them from risky procedures or violations of their privacy by parents eager to collect reproductive material from them.

Hence, although the right to an open future as formulated by Feinberg and applied by Davis has been used in promoting fertility preservation for children, it does not straightforwardly transfer well to this discussion and/or might require, at least in some cases, something more than parental intervention.

[3] I am grateful to Guido Pennings for raising this point during the Q&A following my talk on this topic at the Maastricht symposium "Parental responsibility in the context of neuroscience and genetics".

12.4 Implications of a Parental Obligation to Preserve Children's Fertility

The prospect of aggressive treatments and the general background of medical conditions that threaten children's reproductive potential, make decisions about whether to try to preserve reproductive material from children particularly difficult and very often urgent. If parents should be allowed, encouraged, or if it is part of their parental responsibility to take fertility preservation measures for their children because of certain conditions or treatments, then perhaps all parents should be allowed, encouraged, or are responsible to preserve their children's reproductive potential even in the absence of any immediate threat – and indeed, especially in the absence of such a threat. All children are at risk of not being able to fulfil their reproductive wishes in the way in which they wish, once they reach adulthood. Some postpone reproduction until it becomes difficult to achieve. Some of the reasons behind this postponement may have a lot to do with the family they were was born into and the educational and social status opportunities that they had access to – and some with the current organisation of life, with the need to spend more years in school and work more years before securing a stable income and a home. Parents therefore have reasons to work towards improving societal conditions that currently encourage postponement of reproduction. Furthermore, European demography data suggests that, after years of decline, support for parents encourages higher fertility rates (European Commission et al. 2011: 2). If children have a right to fertility preservation, or otherwise parents ought to try to preserve their children's fertility, then if this is true, parents should aim to ensure a supportive environment for (prospective) parents in their societies.

The causes of infertility are often genetic: they are a part of one's family's genetic history. For example, a man with Y-chromosome deletions may father sons who are infertile, and women who are fragile X syndrome carriers are at risk of premature ovarian failure because of this (Layman 2003). Thus whether, when, and how often one will reproduce has a lot to do with one's parents, either socially or genetically. Moreover, when children are still young, parents *can* decide to try to preserve their reproductive potential. Indeed, as the example of children facing fertility affecting treatment shows, it is sometimes argued that they have a duty to try, and that children have a corresponding right to their fertility being rescued. It could be argued here that the idea of parental responsibility for children *as adults* is less compelling than in the case of children whose reproductive potential is under immediate threat, *while* they are in their parents' care. However, inasmuch as some outcomes are determined by factors largely outside the adults' control (such as their genetic make-up or the social circumstances in which they grow up), and the parents *could have* acted to relieve the impact on fertility of these circumstances, *and* if they have a responsibility to rescue their children's fertility when it is in immediate threat, then parents also have the responsibility to act on these other factors. The very aim of fertility preservation measures is to safeguard an open future for children *for when they are adults*. There is little benefit for children *as children* that their reproductive potential is preserved. I will come back to this shortly.

There are many reasons why, if parents have a duty to try to rescue the fertility of their children suffering from conditions or facing treatments which could compromise their possibility to reproduce genetically, then all parents have this duty for all of their children. Children can lose their reproductive potential in many ways, and indeed the quality of their reproductive material starts to be compromised from their early teenage years (Hens 2017); moreover, as mentioned above, their reproductive desires can also be frustrated in non-medical ways. There are reasons to recommend that parents do this before knowledge of any disease or syndrome which might jeopardise their children's fertility (clearly this may not apply to risky, invasive, or experimental procedures such as ovarian tissue preservation). First, the context of the decision is not fraught with the fear of one's child's life being in immediate danger, nor with the information that the child is likely to become infertile soon. Free from such pressures, parents can calmly consider all the pros and cons and any intervention on the child could be carried out when the child is otherwise safe and sound. Arguably, any (parental) decision is more autonomous the freer it is from immediate fears or worries and the less it immediately follows dramatic life-changing news that requires immediate action. Second, should the parents decide to proceed with fertility preservation when their children are fine and healthy, any procedure taken to that end could take place in the absence of immediate threats and of fragile state due to disease. Furthermore, a child that appears healthy today could be facing a treatment or a condition that threatens or takes away their fertility. So taking action offers insurance, from the point of view of the child's reproductive future, before knowledge of such possible futures comes into play and puts pressure on any decisions.

Should we conclude that parents are entitled, or even have an obligation to preserve their children's fertility, regardless of whether and maybe especially when the children are not in any danger to their life or fertility, there remains the question of what age is appropriate for such an intervention. As I mentioned above, fertility preservation measures such as ovarian cryopreservation have been taken on children as young as two. At that age, the child is too young to understand the purpose of the procedure, which is invasive and offers only a small chance of fertility preservation. It could here be argued that such cases (at least) are those that set apart the strength of the case for fertility preservation when fertility is acutely threatened versus children for whom perhaps fertility preservation measures could be taken when they are older. If, however, a technology were available, or could be developed, that would enable parents to have reproductive material from their children collected already at the time of birth, and stored for future use by the child as an adult, perhaps that is what parents, and scientists, should aim for.

In the midst of technological promises and concerns for children's future, one obvious aspect that has to be kept explicit is that children *qua children* are not going to benefit from the goods of fertility preservation. The effort is undertaken for the sake of a desire to reproduce that they might develop when they are adults. This is important at least in two respects. First, as children grow, they can (and should) increasingly be involved in decisions that concern them – and indeed the thought of a protesting child being in some way made to allow tissue or gametes to be collected from her for preservation is disturbing. However, if they have this right, then perhaps they should be persuaded to allow the procedure, or it should be performed without their knowledge. Many adults may have wanted to become parents since they were children themselves, but many have not. Why *should* the adult have to pay for the lack of an interest in reproduction that she demonstrated as a child? And when she is an adult with a desire to reproduce, is she responsible for her lack of fertility because she refused fertility preservation as a child, or is she equally deserving as children who have assented?

Second, fertility preservation is undertaken for the sake of the adult that the child will become – and not for her sake as a child. A child who faces chemotherapy and radiation treatment is already missing out on at least some of the goods of childhood for an indeterminate duration, and her survival is threatened. Maybe she will not become an adult. Should this child be subjected to even the slightest risk to her wellbeing in order to allow a possibility that she may be interested in as an adult? The answers to these questions are not obvious – and it is equally not obvious that intervention on the children for the purpose of fertility preservation is permissible at all, let alone a duty.[4]

12.5 Infertility, Desert, and Fairness

It is easy to understand why one might consider preventable infertility, caused by medical or genetic conditions or medical treatments, to be "doubly unjust" and "unfair". However, this consideration relies on an assumption that other avenues towards incapacity to reproduce are less unjust or unfair. Thus arguments from justice and fairness here function on an assumption of *desert*: adults who have lost their fertility as children because of disease have a different type of claim (or *deserve* the treatment more) than those who may end up needing assistance for other reasons. This claim is problematic in several ways. The reasons why people cannot reproduce can vary significantly in each particular case, and can be very personal. For one person, it might be because of her choice of partner: because her partner is of the same sex as herself or is infertile. People are not usually expected to simply change partner when they discover their partner's infertility: on the contrary, in the Western world that would be seen as shallow, uncommitted, disrespectful and unloving towards that partner. [On the other hand, same-sex partners *may be* expected to simply disregard their sexual orientation and seek a partner with whom they are reproductively compatible (Pennings 1999; Smajdor and Cutas 2015)]. How is this person, in this circumstance, responsible for her incapacity to reproduce, and in general why is this not unfair and unjust when compared with people who did partner with someone who proved to be fertile, or reproductively compatible with her?

---

[4] I thank my co-editor Kristien Hens for inspiring these concerns.

Someone else might have a harder time finding a partner because of the environment she was born in: such as a small isolated community, or a minority ethnicity or immigrant background; or because she does not fit standard understandings of attractiveness of her time and location. It would be hard to construct an argument for parents' responsibility for their child's likelihood of finding a reproductive and parenting partner. However, in such cases as well as in that of medical or genetic conditions that threaten fertility, there *are* things that are in the parents' power to do before their children's reproductive opportunities are compromised. Furthermore, according to recent findings, activities such as smoking can start to alter a boy's reproductive potential already from puberty (Hens 2017). This is one more reason for parents to do a service to their sons (and grandchildren) by removing and preserving reproductive material from them before it is altered. Therefore, perhaps parents *should* act to preserve their children's fertility.

Young people are not always very good at judging the risks of postponing reproduction (Lampic et al. 2006; Schmidt 2010), and the way that society is structured in the Western world very much encourages them to postpone: by the time they finish their education and find a stable workplace and can afford a home, their fertility (especially women's) will be spiralling down. Throughout Europe, the highest unemployment rate is in the under 25s, and employment precariousness particularly affects women (Eurostat 2014). When they are employed (again, especially women, because they carry the pregnancy and take the most of parental leave), their employers are likely to explicitly encourage them to postpone indefinitely – and sometimes even offer support for fertility preservation (BBC News 2014). Young adults are not responsible for these circumstances, and it would hardly be responsible to expect them to reproduce at the peak of their fertility, thus limiting their own futures and producing children with a more precarious start in life than they would if they had waited. But if they wait, then their chances of reproducing will decrease. That they have to make such choices may also not be fair: whatever they do, they jeopardise some (education, employment, economic security) or other (fertility) parts of their future. Thus, if children facing fertility threatening treatment *deserve* fertility preservation, so do other children, and if parents should take steps to preserve the fertility of their children who face immediate risks, perhaps so too should parents of children who later on could face more diffuse but equally serious risks.

Lastly, one of the comments cited above (Larcher 2012), supporting a distinction between the case for access to ARTs of adults who are rendered infertile because of "necessary treatment" and that of, presumably, those who avail themselves of ARTs because of other reasons, suggests an undercurrent of implicit assumptions. The author argues that imposing costs for ARTs on those who are infertile because of necessary treatment is unfair to them. While this may be the case, it may be equally unfair for those who are infertile for other reasons. Presumably it is possible for someone to want to access ARTs through some fault of their own and/or without having any need for it which is at least as valid as that of a childhood cancer survivor. Even so, it is difficult to envisage a fair scheme for support of access to ARTs in terms of responsibility for one's infertility: is the partner of an infertile person responsible for her incapacity to reproduce? How about a gay man or a lesbian woman? Or someone who could not find a partner with whom to reproduce? And so on: if reproductive potential and reproduction are a right or a need, then it is hard to make a case for why those who suffer from infertility because of necessary treatment as children should have this right and this need acknowledged and supported, and others should not.

12.6 Fertility Preservation Doth Not a Parent Make

Clearly, collecting and preserving reproductive material from and for one's child has some potential to contribute to the array of choices that she will have as a grown-up. However, there is no relation of implication between having one's genetic material collected and stored as a child and being able to reproduce genetically. Firstly, once the reproductive material is removed and stored, one does not simply have access to it. Depending on the legislature, the now grown up child may not have access to it because she is single, too old, in a same-sex relationship, or has an infertile partner; or she may not afford the treatment (in some countries the costs of fertility treatments are not covered at all from health insurance, and in some countries only some treatments or only a certain number of IVF cycles are covered, and/or only in some cases). Thus in their pursuit to keeping their children's futures open, if parents should preserve their children's fertility, then they should also contribute to eliminating such other, social or legal, barriers. For example, Italy has very strict regulations for the use of ARTs, which are only allowed for "adult couples of different sex, married or cohabiting, of reproductive age, both living" (Art. 5 Law no. 40, 19 February 2004, my translation), and only with their own gametes (Art. 4). Thus, for the childhood cancer survivors in Italy whose reproductive material has been collected and preserved by their parents and could be used in ARTs, they can only have access to it within a certain timeframe, and only if and while they live with or are married to a person of a different sex, also of reproductive age, who does not happen to lack gametes of her own. However, when this restrictive law was subjected to a national referendum in 2005, less than 26 % of Italian citizens chose to exercise their right to vote. This choice to not repeal the law has made reproduction much more difficult or altogether out of reach for many – some of whom may be the children of people who chose to not vote. If parents have a responsibility to contribute to their children having more rather than less reproductive choice in the future, then they also have a responsibility to repeal restrictive regulations such as the Italian law governing ARTs: at the very least, because such regulations could simply render any attempts to preserve fertility useless because the children will not have access to their own reproductive material.

Secondly, even if all other elements align, if the genetic material can be used and at adulthood the child wants to use it and is allowed to do so, the chances of her coming out of this process with a live birth are limited: and the more experimental the procedure, the lower the probability that she will become a parent using the material preserved by her parents. Even as IVF success rates are rising steadily, along with an increase in successful results from fertility preservation for children, this is progress that requires major investments and offers no guarantees.

At the same time, it is unclear what the landscape of fertility treatments will look like in a decade or two. Mice germ cells have been obtained from skin cells, and these have led to the creation of offspring (Hayashi et al. 2011, 2012). Research towards obtaining human gametes in vitro from other types of cells is ongoing in several labs around the world (Chuva de Sousa Lopes and Roelen 2015). Should this work lead to treatments for humans in a decade or two, the efforts and costs of fertility preservation for children might prove to have been unnecessary even if the child does develop a wish to reproduce.

Another possibility that some families might consider in their quest to preserve their children's chances to have genetically related children is to insure that gametes from someone in the family are available by the time they become adults. For example, parents could store their own genetic material for future use by their children; or a sibling could be encouraged to consider intra-familial gamete donation. This could give children a chance to have offspring who are genetically related to them – even if not technically their genetic offspring. Moreover, this could allow them to be both genetically related and the biological parents of the resulting offspring, by the grafting of familial testicular tissue in the case of men, or of ovarian tissue in the case of women, and/or by women carrying the resulting pregnancy (either because the eggs mature in their bodies, or via IVF and embryo transfer). The desire to reproduce is often made of a mix of ingredients that are not the same for all (Smajdor and Cutas 2015). For example, some years ago a woman preferred to have an ovarian tissue transplant rather than eggs from her sister, because she wanted the eggs to mature in her own body (BBC 2007). Genetic relatedness *and* begetting *in the natural way* might thus come closer to fulfilling the desires of some survivors than having current fertility preservation measures undertaken on them as children.

Having been subjected to invasive surgery as a child might make the grownup attribute anticipatory importance to the genetic link: how can she not see this as important when so much effort and so many resources have been put into keeping this option open for her? Can this knowledge *plant* a desire to reproduce in her as she grows up that would not otherwise have formed, along with the expectation that this stored material holds the key for her fertility? And *if* this happens, then the frustration of that desire that could come from failure to reproduce with that material might just be an example of unnecessary and avoidable suffering, caused by others deciding that this path be taken for her. In taking this path on behalf of a child, in the words of my co-editor Kristien Hens, her life story is changed. This issue of the risk of fertility preservation translating into expectations and pressure to reproduce has been raised by other authors as well (Smajdor 2007; Quinn et al. 2012; Satkoske and Parker 2013; Cutas and Hens 2015).

One other option that parents have to prevent their children from suffering because of infertility as adults is that instead of subjecting them to fertility preservation techniques, they foresee and treat the reproductive desire instead.[5] They could do this by impressing upon their children from an early age the values of parenting other than genetic filiation, and by changing the environment around the children to emphasise these values over genetic relatedness. Research on families in which there is, and in which there is not a genetic link between parents and children, indicates that what seems to matter most for children's wellbeing is family functioning (the quality of the relationships within the family) rather than whether parents and children share a genetic link (Golombok 1998, 2000, 2015; Golombok et al. 1995; Bornstein 2002; Hastings et al. 2006; Biblarz and Stacey 2010; Lamb 2012; Scheib and Hastings 2012). Such findings are not always in line with popular beliefs about families – and cancer survivors considering parenting, as well as anyone else, could benefit from alternative interpretations of "blood is thicker than water" ("love is thicker than blood" is one such possible interpretation). This is important, because not being able to reproduce genetically does not equal not being able to become a parent – and a chance of genetic parentage is not the only way in which children's futures can be enriched.

12.7 Conclusion

In this chapter, I have examined whether parents have a duty to preserve the fertility of their children facing fertility impairing treatments. I have presented how the children's "right to an open future" is said to generate this duty. I have put these cases in the context of parents and children in general, and I provided some arguments for why, if children's right to an open future generates their parents' duty to take measures to preserve their reproductive potential, then when possible this duty can and perhaps should be discharged before the parents acquire knowledge of any acute dangers to this potential. I have also discussed some of the elements that are said to set apart infertility caused by disease or its treatment on the one hand, and infertility caused by other factors, such as postponement of reproduction, or choice of partner, on the other hand: in particular, arguments in terms of desert and fairness. I have then looked at the long, uncertain and broken road from preserving children's fertility to their becoming parents.

Fertility preservation for children is discussed in a context in which parents are confronted with major decisions regarding their children's life. The argument for fertility preservation in such situations is for them to take possibly costly and risky steps at a very dramatic time for the sake of a desire that their children may or may

[5] As my co-editor Dorothee Horstkötter pointed out, such a proposal "might also be tricky, because one could turn the argument around and say that 'foreseeing and treating reproductive desires' should entail a clear discouragement from engaging in a same-sex relationship, or – for women - in higher education and employment as that facilitates postponement." This may be true, but by doing so they would restrict their children's futures in other, more predictably harmful ways – and thus fail *more*, and more actively, at safeguarding their children's interests than they would by not taking fertility preservation measures on their behalf.

not form (to reproduce) and that they subsequently may or may not be allowed or able to fulfil. Different people's reproductive desires have different contents, and these are sensitive to the values endorsed in the cultures to which they are exposed. Alongside reflecting on the prospect of fertility preservation on behalf of their children, parents should also consider other ways in which they can keep their children's future open: for example, by advocating or voting against restrictive regulations that might one day prevent their children from accessing fertility treatments or becoming parents, as well as by contributing to make societal conditions more friendly to uptake of parenting – should one develop the wish to become a parent.

I will leave the question of whether children do have a right to fertility preservation, as well as whether their parents (or other entities) have a corresponding duty, open. What I hope to have achieved in this chapter is to raise a series of issues for consideration when pondering this question and to show some of the possible implications of a duty to keep children's reproductive futures open.

References

BBC News, Ovarian transplant first welcomed. August 2, 2007. Available at news.bbc.co.uk/2/hi/health/6924014.stm. Accessed September 2016.

BBC News, Tech giants to freeze eggs for their female employees. October 15, 2014. Available at www.bbc.com/news/business-29626291. Accessed September 2016.

Biblarz, T., and J. Stacey. 2010. How does the gender of parents matter? *Journal of Marriage and Family* 72(1): 3–22.

Bizet, P., et al. 2012. Sperm cryopreservation before cancer treatment: A 15-year monocentric experience. *Reproductive Biomedicine Online* 24(3): 321–330.

Bornstein, M.H. (ed.). 2002. *Handbook of parenting: Vol. 1: Children and parenting*. Mahwah: Erlbaum.

Bourke, E., et al. 2014. A qualitative exploration of mothers' and fathers' experiences of having a child with Klinefelter syndrome and the process of reaching this diagnosis. *European Journal of Human Genetics* 22(18): 18–24.

Chuva de Sousa Lopes, S.M., and B.A.J. Roelen. 2015. Current status of *in vitro* differentiation of stem cells into gametes. *Animal Reproduction* 12(1): 46–51.

Cutas, D., and K. Hens. 2015. Preserving children's fertility: Two tales about children's right to an open future and the margins of parental obligations. *Medicine, Healthcare and Philosophy* 18(2): 253–260.

Davis, D.S. 1997. Genetic dilemmas and the child's right to an open future. *Hastings Center Report* 27(2): 7–15.

Demeestere, I., et al. 2015. Live birth after autograft of ovarian tissue cryopreserved during childhood. *Human Reproduction* 30(9): 2107–2109.

Dittrich, R., et al. 2012. Live birth after ovarian tissue autotransplantation following overnight transportation before cryopreservation. *Fertility and Sterility* 97(2): 387–390.

Dolin, G., D.E. Roberts, L.M. Rodriguez, and T.K. Woodruff. 2010. Medical hope, legal pitfalls: Potential legal issues in the emerging field of oncofertility. *Oncofertility. Cancer Treatment and Research* 156: 111–134.

European Commission, Directorate-General for Employment, Social Affairs and Inclusion & Eurostat 2011. *Demography report. (2010),* Commission Staff Working Paper. Available at http://epp.eurostat.ec.europa.eu.

Eurostat. 2014. *Unemployment statistics.* Available at http://ec.europa.eu/eurostat/statistics-explained/index.php/Unemployment_statistics. Accessed September 2016.

Fallat, M.E., and J. Hutter. 2008. Preservation of fertility in pediatric and adolescent patients with cancer. *Pediatrics* 121(5): e1461–e1469.

Feinberg, J. 1980. The child's right to an open future. In *Whose child? Children's rights, parental authority, and state power*, ed. W. Aiken and H. LaFollette. Totowa: Rowman and Littlefield.

Feinberg, J. 1992. *Freedom and fulfilment: Philosophical essays.* Princeton: Princeton University Press.

Fujita, K., et al. 2005. Transplantation of spermatogonial stem cells isolated from leukemic mice restores fertility without inducing leukemia. *The Journal of Clinical Investigation* 115(7): 1855–1861.

Gies, I., et al. 2012. Spermatogonial stem cell preservation in boys with Klinefelter syndrome: To bank or not to bank, that's the question. *Fertility and Sterility* 98(2): 284–289.

Golombok, S. 1998. New families, old values: Considerations regarding the welfare of the child. *Human Reproduction* 13: 2342–2347.

Golombok, S. 2000. *Parenting: What really counts?* London: Routledge.

Golombok, S. 2015. *Modern families. Parents and children in new family forms.* Cambridge: Cambridge University Press.

Golombok, S., et al. 1995. Families created by the new reproductive technologies: Quality of parenting and social and emotional development of the children. *Child Development* 66: 285–298.

Grundy, R., et al. 2001a. Fertility preservation for children treated for cancer (1): Scientific advances and research dilemmas. *Archives of Disease in Childhood* 84(4): 355–359.

Grundy, R., et al. 2001b. Fertility preservation for children treated for cancer (2): Ethics of consent for gamete storage and experimentation. *Archives of Disease in Childhood* 84(4): 360–362.

Hastings, P., et al. 2006. *Children's development of social competence across family types.* At www.samesexmarriage.ca/docs/Justice_Child_Development.pdf. Accessed September 2016.

Hayashi, K., et al. 2011. Reconstitution of the mouse germ cell specification pathway in culture by pluripotent stem cells. *Cell* 146(4): 519–532.

Hayashi, K., et al. 2012. Offspring from oocytes derived from in vitro primordial germ cell-like cells in mice. *Science* 338(6109): 971–975.

Hens, K. 2017. The ethics of postponed fatherhood. *International Journal of Feminist Approaches to Bioethics* 10(1): 1.

Herlihy, A.S., et al. 2011. Postnatal screening for Klinefelter syndrome: Is there a rationale? *Acta Paediatrica* 100(6): 923–933.

Imbert, R., et al. 2014. Safety and usefulness of cryopreservation of ovarian tissue to preserve fertility: A 12-year retrospective analysis. *Human Reproduction* 29(9): 1931–1940.

Jadoul, P., et al. 2010. Fertility preservation in girls during childhood: Is it feasible, efficient and safe and to whom should it be proposed? *Human Reproduction Update* 16(6): 617–630.

Jahnukainen, K., et al. 2006. Clinical potential and putative risks of fertility preservation in children utilizing gonadal tissue or germline stem cells. *Pediatric Research* 59(4 Pt 2): 40R–47R.

Kelleher, S., et al. 2001. Long-term outcomes of elective human sperm cryostorage. *Human Reproduction* 16(12): 2632–2639.

Lamb, M. 2012. Mothers, fathers, families, and circumstances: Factors affecting children's adjustment. *Applied Developmental Science* 16(2): 98–111.

Lampic, C., A.S. Svanberg, P. Karlström, and T. Tyden. 2006. Fertility awareness, intentions concerning childbearing, and attitudes towards parenthood among female and male academics. *Human Reproduction* 21(2): 558–564.

Larcher, V. 2012. The ethical obligation to preserve fertility in the face of all therapies that might adversely affect it. *Archives of Disease in Childhood* 97(9): 767–768.

Layman, L.C. 2003. Genetic causes of human infertility. *Endocrinology and Metabolism Clinics of North America* 32(3): 549–572.

Legge 40, 19 February 2004, available at www.camera.it/parlam/leggi/04040l.htm.

Magelssen, H., et al. 2005. Twenty years experience with semen cryopreservation in testicular cancer patients: Who needs it? *European Urology* 48(5): 779–785.

Mehta, A., and D.A. Paduch. 2012. Klinefelter syndrome: An argument for early aggressive hormonal and fertility management. *Fertility and Sterility* 98(2): 274–283.

Nisker, J., F. Baylis, and C. McLeod. 2006. Choice in fertility preservation in girls and adolescent women with cancer. *Cancer* 107(7): 1606–1689.

Pennings, G. 1999. Measuring the welfare of the child: In search of the appropriate evaluation principle. *Human Reproduction* 14(5): 1146–1150.

Poirot, C., and B. Schubert. 2011. Fertility preservation in prepubertal children. *Bulletin du Cancer* 98(5): 489–499.

Quinn, G.P., et al. 2012. Preserving the right to future children: An ethical case analysis. *The American Journal of Bioethics* 12(6): 38–43.

Rieker, P.P., et al. 1990. Adaptive behavioral responses to potential infertility among survivors of testis cancer. *Journal of Clinical Oncology* 8(2): 347–355.

Rochman, B. 2011. *Even as babies, cancer patients strive to preserve fertility*. http://healthland.time.com/2011/03/22/children-having-children-when-a-kids-got-cancer-fertility-preservation-can-help/. Last Accessed September 2016.

Saito, K., et al. 2005. Sperm cryopreservation before cancer chemotherapy helps in the emotional battle against cancer. *Cancer* 104(3): 521–524.

Satkoske, V.B., and L.S. Parker. 2013. Preserving testicular tissue and a boy's open reproductive future. *The American Journal of Bioethics* 13(3): 36–37.

Scheib, J., and P. Hastings. 2012. Donor-conceived children raised by lesbian couples: Socialization and development in a new form of planned family. In *Families – Beyond the nuclear ideal*, ed. D. Cutas and S. Chan. London: Bloomsbury Academic.

Schmidt, L. 2010. Should men and women be encouraged to start childbearing at a younger age? *Expert Reviews in Obstetrics Gynaecology* 5(2): 145–147.

Schover, L.R. 2005. Motivation for parenthood after cancer: A review. *Journal of the National Cancer Institute Monographs* 34: 2–5.

Simon, B., et al. 2005. Preserving fertility after cancer. *CA: A Cancer Journal for Clinicians* 55(4): 211–228.

Smajdor, A. 2007. There's no such thing as a free egg. *BioNews*. Available at www.bionews.org.uk/page_37947.asp. Accessed September 2016.

Smajdor, A., and D. Cutas. 2015. Will artificial gametes end infertility? *Health Care Analysis* 23: 134–147.

Thomson, A.B., et al. 2002. Semen quality and spermatozoal DNA integrity in survivors of childhood cancer: A case–control study. *Lancet* 360(9330): 361–367.

van Casteren, N.J., et al. 2008. Use rate and assisted reproduction technologies outcome of cryopreserved semen from 629 cancer patients. *Fertility and Sterility* 90(6): 2245–2250.

Chapter 13
Family Responsibilities and Genetic Disorders in Yoruba Culture: The Example of Sickle Cell Anaemia

Ademola Kazeem Fayemi

Abstract This chapter discusses the responsibilities of family caregivers in the evolving process of genetic biotechnology in Yoruba-African culture. Using sickle cell anaemia (SCA) as an example of paediatric genetic disorder, I argue that there are tensions between traditional family care structures, widespread cultural beliefs about genetic conditions and the possibilities in modern genetics and medicine. Such tensions affect the expectations of responsibilities toward children with SCA. In addressing such tensions, I make a case for the integration of new biomedical opportunities with traditional Yoruba childcare ethics in ways that promote the best care for children with SCA.

Keywords Family caregivers • Paediatric genetic disorder • Yoruba culture • Sickle cell anaemia • Caregiver's responsibilities • Personhood

13.1 Introduction

Genetic testing and diagnosis of genetic disorders have not been available in Nigeria until very recently when the Chevron/Nigerian National Petroleum Corporation (NNPC) Joint Venture donated money to establish an ultramodern Molecular Biology Laboratory Centre (MBLC) at Lagos University Teaching Hospital, Lagos, in February 2014. The main task of the Centre is to improve healthcare by providing DNA biotechnology services such as prenatal diagnosis of sickle cell anaemia, neonatal genetic testing, new-born screening, cancer screening parental genetic testing, and other genome related screenings. Before this time, genetic testing and diagnosis of various forms (parental testing, prenatal screening, pre-implantation genetic diagnosis (PGD), etc.) were only conducted overseas. With the functional operation of MBLC beginning over a year ago, its capacity for diagnosing mutations linked to

A.K. Fayemi (✉)
Department of Philosophy, University of Lagos, Lagos, Nigeria
e-mail: kcaristotle@yahoo.com

K. Hens et al. (eds.), *Parental Responsibility in the Context of Neuroscience and Genetics*, International Library of Ethics, Law, and the New Medicine 69,
DOI 10.1007/978-3-319-42834-5_13

different kinds of genetic disorders and improving the treatment of patients of diverse socio-economic classes in Nigeria is significant. This development promises to reduce the incidence of medical tourism by Nigerians who seek genetic testing or diagnosis services. Moreover, the forensic testing offered by MBLC is important, as well, for legal and justice administration in Nigeria.

While MBLC has been welcomed by Akin Osibogun (2014), the Chief Medical Director of the University of Lagos Teaching Hospital, as a positive development in the Nigerian health sector, it raises a host of challenges for bioethics in Nigeria. New moral problems and dilemmas will arise in the specific cosmopolitan cultural context that MBLC is located in, which should be identified and addressed. One of these challenges is the potential implication(s) of the genetic services and interventions of MBLC for the traditional Nigerian culture in terms of the responsibilities of family caregivers. In Nigeria, it is commonly assumed, without critical questioning, that there is symbiotic and complementary work division between the care a patient receives from the professional healthcare team in the hospital and that received from the patient's family. However, in the atmosphere of emerging genetic biotechnology in Lagos, it is becoming increasingly difficult to ignore the imperative of addressing the question of what the responsibilities of family caregivers are. It may be also be asked where such responsibilities end and where those of healthcare professionals begin in the era of genetic technology.

In situating the foregoing into a specific context, I will use the example of paediatric genetic disorder. The reason for this choice is the fact that many Nigerian children have become more vulnerable to inherited genetic diseases than children in the biotechnologically advanced world. They are more vulnerable due to recurring cases of misdiagnosis resulting from lack of genetic centres, of genetic counselling, genetic testing or diagnosis of adult-onset mutations and appropriate care management. More fundamentally, such vulnerability is heightened by tension between widespread cultural beliefs and perceptions about genetic conditions and the orthodox medical perception. These different perceptions often result in conflicting views and expectations regarding which of these parties have which responsibilities.

In order to illustrate an instance of such tension, this chapter aims at discussing family responsibilities in the Yoruba culture using the case of sickle cell anaemia (SCA) as an example. Though the tension between traditional practices, beliefs and a modern, scientific model, is not unique to the Yoruba experience, it is worth pointing out how such tension creates a moral dilemma in the context of paediatric care. The Yoruba people are one of the major ethnic groups in southwest Nigeria. Contextualizing the discussion on the Yoruba is very meaningful because the MBLC is located in a dominantly Yoruba area of Lagos State. It is also illustrative because autosomal recessive diseases like sickle cell anaemia affects about 1–3 % of the Nigerian population (Okere et al. 2014). Besides the fact that SCA is one of the most common genetic disorders with a high frequency of genotype (HbsHbs) among the Yoruba (Taiwo et al. 2011), the prevalence of infant morbidity and mortality from SCA in Nigeria is relatively high (WHO 2006). Nigeria is a state where genetic technology, counselling, medical insurance, standard care for managing and caring for sickle cell patients, as known in the contemporary Western world, are still at a low ebb.

On account of all the aforementioned factors it is important to understand the emerging responsibilities and obligations of family caregivers to children with SCA among the Yoruba and their justifications. This chapter seeks, therefore, to discuss interrelated questions: Should contemporary Yoruba re-evaluate their responsibilities in caring for children with SCA in light of the options afforded by the new genetic technology? Should traditional Yoruba beliefs and ethics of children's care be given priority over and above what some may call the responsibility to avoid the birth of children with certain genetic conditions? How do we responsibly negotiate and integrate the possibilities of new biomedical opportunities with traditional Yoruba beliefs and ethics in a way that promotes the best care for children with SCA?

13.2 Sickle Cell Anaemia in Yoruba Culture

SCA is an autosomal recessive disease which exhibits a range of severity (Ohaeri and Shokunbi 2002). It is a condition that "alters the shape of the red blood cells from round to 'sickle' shape, causing them to block small blood vessels and interfere with normal blood flow" (Okere et al. 2014, 87). Typical symptoms of SCA include chronic incidents of pain in the body joints, chest and abdomen. SCA is a potentially life-threatening condition that can lead to bacterial infections as well as organ failure. The disease is at this time incurable, and those who suffer from it have an average life expectancy of less than 50 years (Okere et al. 2014).

Among the Yoruba, *abiku* (born to die prematurely) is a general word used to refer to persons suffering from SCA. The Yoruba are generally aware of the chronic and endemic nature of the manifestations of this disease. The way caregivers in Yoruba culture understand and interpret symptoms of diseases generally and genetic disorders in particular, has a critical impact not only on the ways in which they understand the responsibilities that they have towards the patients, but also on the type of diagnosis and treatments that they apply. Traditionally, the Yoruba believed that the source and manifestations of the SCA crisis are paranormally constituted, defying scientific explanations.

There are many genetic conditions that are attributed to the will of the gods in traditional Yoruba culture due to lack of scientific explanation: albinism, Down syndrome, and sickle cell among others. These health challenges, whether in children or adults, are understood to be caused by supernatural factors such as predestination, being a punishment or curse by enemies or ancestral spirits, and sometimes they are understood as 'disguised blessings' from the gods and divinities. SCA is interpreted as an effect of a spiritual affliction or predestination, especially if the patient has completed treatment with herbal anti-sickling preparations without lasting improvements in her health. Traditional herbal medicine for SCA among the

Yoruba consists of *E. caryophyllata*, *P. guineense*, and *P. Osun*, extracted with *ogoro* (ethanolic distillate of palm wine) (Ameh et al. 2012).

At the same time, there is a low level of knowledge, awareness and clinical diagnosis of many genetic disorders in Yoruba traditional culture. Even in contemporary times where there are improved clinical studies of herbs in the treatment of ailments and disorders generally, some issues still hamper awareness of genetics among the Yoruba. Such challenges include the unavailability of behavioural interventions, of adequate genetic research centres, genetic counsellors and trained geneticists relative to the population, and unaffordable treatment costs. This does not mean that there is no condition which is correctly understood as a genetic disorder in Yoruba culture. Dwarfism, for instance, is clearly expressed in many popular proverbs as a genetic condition.

13.3 Family Responsibilities for Children in Traditional Yoruba Culture

One of the key components of acculturation in several African societies is responsibility. Responsibility takes different dimensions ranging from being aware, having a predisposition to act and the actual performance of such responsibility. In Yoruba culture, the communal norms and precepts of the society require that each and every one sees the meaningfulness of each in the wellbeing of the other. The society is organized on extended family structures, a network of kinship, where biological parenthood plays a critical but not exclusive role in the raising of children. While the Yoruba do not deny the complementary roles expected of a father and a mother in childrearing, they believe that the support of both the parents and the extended family is essential in caring for children. Children are seen as 'ours' and not 'mine' (Fayemi 2014). The individualistic and exclusive elements of the nuclear family are absent in the raising of children among the Yoruba, and because of this, responsibilities are not merely parental, but collectively shared and shouldered among members of the extended family who collectively constitute the caregiver.

Caregivers' responsibilities to children in traditional Yoruba culture begin at the prenatal stage. A pregnant woman receives a level of care comparable to that of a baby. She is cared for and protected by all members of the extended family and not allowed to engage in strenuous tasks and jobs. However, no one prevents her from engaging in them, if she wants to. Today, the often harsh economic situation in contemporary Yoruba societies is beginning to change this protective attitude towards pregnant women. The caregivers have the responsibility of educating and guiding expectant mothers in avoiding prenatal substance exposure and maternal consumption of foods, alcohol and drugs that can have adverse health outcomes for her and her pregnancy. Acting as *de facto* nurses and medication dispensers, family members work closely with the family physician in ensuring safe labour and delivery.

Segun Gbadegesin eloquently underscores this point:

> A new baby arrives into the waiting hands of the elder members of the household. From that point on, they [caregivers] see the baby as theirs [but also her mother's]. All members of the community invest their time and resources on her [the expectant mother]...The process of socialization begins right from birth and all extended family members have a role to play. The structure of the family compound makes this easy since everyone is close by. A child cannot misbehave without being corrected immediately. Love is lavished, but the rod is not spared. In this kind of environment, growing children are able to see themselves as a part of an extended family household and not as atoms (Gbadegesin 2013: 7).

An authoritative Yoruba proverb on child nurturing says *oju merin lon bi'mo, egberun oju ko to woo* (children are biologically born by four eyes, those of the mother and the father, but collectively nurtured by more than a thousand eyes). Such an organization of care among the Yoruba involves both men and women and is not solely the concern of the biological parents. All adult members of the family, uncles, aunts, and grandparents, both paternal and maternal, including the parents, are responsible for the care and nurture of children. For this reason, the responsibility to care for children is not strictly that of the biological parent; even where parents neglect their responsibilities to their children, other family members make up for the gap. This situation is reflected in the Yoruba proverbial saying that *obini ko lo lomo bi kose on' wooni* (authority and responsibility to a child are not only a function of biological parenthood, but are equally determined by caregivers who nurture the child).

Caregivers' responsibilities are holistic and continuous. They are holistic in the sense that the caregivers are responsible for providing the physical, psychological, emotional, economic, spiritual, health and educational security for the children in the family clan. Among the Yoruba, caregivers' responsibilities to children are a lifelong commitment because regardless of how old the children have become, they are still children to their caregivers. It is only the content of these responsibilities that changes over time as children grow into adulthood.

At infancy, all family members are expected to care for all the children in the family compound. The only specific duty of the biological mother at this point is to breastfeed the baby. Other care-related actions and responsibilities (cuddling, back-carrying the baby, nurturing, keeping the child healthy and chanting of lineage and personal *oriki* (praise poetry)) are performed on her behalf by the members of the family. While the teenagers in the family cuddle, nurture the infant and carry her on their backs, paternal and maternal uncles, grandparents, and other relatives together with the biological father have the responsibility of ensuring the healthy development of the children through preparation and application of herbal medicines.

Every family has their traditional physician who visits regularly to check on the health of the children and the family members in general. When an ailment defies herbal medications prescribed by the traditional physician, referrals are made by the physician to diviners also known as priests. Diagnosis and treatment of illness by diviners are spiritual and seek the appeasement and containment of the secondary causes of the illness. Keeping the child healthy, which is a key responsibility of the caregivers, is not only a matter of physical stability. The spiritual and psychological wellbeing of the child's whole existence is an important element of responsibility that the caregivers see to. While the diviners are consulted in matters connected with the spiritual welfare of the child, the psychological wellbeing is a duty of all the members of the clan. This is achieved by singing personal praise poetry to children. This act is believed to "evoke a feeling of confidence, pride, wellbeing and love in the child" (Balogun 2012). The child, according to Yoruba culture, is entitled to this care.

As the child develops through childhood to adolescence, the responsibilities of the biological parents as well as the caregivers increase. They play a major role in teaching of values as well as enforcing standards of behaviour among the children. It is their duty to provide leading examples of these standards of behaviour, which the children are to take a cue from. If caregivers fail to nurture and guide children in line with the societal norms and standards, the disgrace extends to the extended family lineage as a whole. In order to avoid the bad reputation the family will earn, serious efforts are usually made by family members to adequately fulfil their responsibility for the welfare of the children (Balogun 2010).

Besides passing cultural norms, taboos and societal precepts on to their children, caregivers are actively involved in the children's education. They have an obligation to help children become educated and productive citizens. The caregivers also have a role in the development of the child from late adolescence to full adulthood. Caregivers are responsible for taking the 'adult child' through the rites of puberty, wedding process and marriage,[1] comforting them in difficult moments and providing guidance based on life experiences.

As children grow into adulthood, caregivers still have the primary responsibilities of ensuring a successful marital home by providing moral support and advice, and by being active in conflict resolution in cases of discord between the couple. At no point in a person's life is the individual isolated from the support of family caregivers. This gives vent to the popular saying that "I am because we are and since we are, therefore I am" (Mbiti 1970: 141). This communitarian attitude involves an obligation to act in order to assist others in a system of mutual obligation, sacrifices and responsibilities. Among the Yoruba, there are mutually shared practices at many levels of social existence – such as childrearing and bereavement.

With regard to children with serious genetic disorder(s), caregivers in the family have moral obligations to care for them in order to enhance their wellbeing. Such

[1]Wedding consists of a long process among the Yoruba. Though the wedding ceremony is the responsibility of the bride's family to organize, the courtship process is instigated by the young man's choice of whom to marry. Upon identification of the prospective bride, the suitor appoints someone in trusteeship (the person is called *alarina*) to convey the message of interest to the woman. Once mutual love is expressed by both parties, their parents are informed. Each family sets up investigative inquiries about the personality, character, family background and health history of the prospective spouse. Once satisfied with the outcome of such inquiry which often involves spiritual consultation as complement of the empirical findings, the man's parents and family head arrange a visitation of the prospective bride's parents and family where their consent is sought. Upon such approval of the relationship, the bride's "price" and gifts are presented to her on an engagement day, which is mutually fixed by both families. The climax of the wedding process is the marriage itself, which is a union of not only the husband and wife, but also of the couple's families.

obligations are expected of caregivers independently of the assumed origin of the disorder. Before resigning to fate in cases of children with SCA, efforts are usually made to improve their capacity for useful living through herbal and spiritual interventions that allow for social integration within the communal network.

13.4 Genetic Disorders and Caregivers' Responsibilities

In contemporary times, many aspects of caregivers' responsibilities in traditional Yoruba society are now dynamically transfusing into new forms of understanding. Changes brought about by the increased availability of technology, acceptance of foreign religions and their values, and other social aspects of modernisation require reanalysis of and perhaps adjustments in the requirements of responsibilities, and a move from the traditional outlook to a modern one. The concept of children's caregivers itself is not unaffected by this modernization process. Caregivers' responsibilities in the traditional context have dwindled in contemporary Yoruba society owing to influences from foreign values, concepts, and practices. The communal social structure and bond based on the solidarity of togetherness and caregivers' responsibilities are beginning to be less of a stronghold, especially in the cosmopolitan cities when compared to the rural Yoruba communities. There are apparent structural, institutional and attitudinal changes from traditional conditions of living to more modern ones, especially in economic, social, political, infrastructure and technological areas (Balogun 2012).

Today, there is the tendency to talk more of *parental* responsibilities than family caregivers' responsibilities, focusing on provision of children's economic, health and financial needs with less input from the holistic care provided by the extended family caregivers. Parental responsibilities to children are now defined in economic terms to the extent that a responsible parent is seen as one meeting the financial obligations and needs of the children (be it in education, health or physical growth) in contemporary Yoruba culture. The holistic care, social security and love of the kinship network made possible through caregivers are fizzling. The introduction of genetic technologies, which open up alternative possibilities for the detection and treatment of SCA, is a further symptom of an increasing Westernisation of Nigerian/Yoruba culture.

Nurturing children through the mechanism of caregivers' responsibilities may face some practical challenges in contemporary times. These challenges notwithstanding, collective arrangements are still relevant in cases where parental care alone is not sufficient (or not possible), and caregivers can fill the vacuum of nurturing and caring for vulnerable children. This is especially important in a non-welfarist state and in developing societies where there are no institutional safety nets for vulnerable children/citizens.

Given the cultural transition of Yoruba society from traditional to modern, it is to be expected that there are shifting loci in the traditionally conceived roles and responsibilities of caregivers to children with SCA. Family caregivers' responsibilities cannot be strictly determined by age-old traditions and beliefs. In order to be able to move beyond the traditional Yoruba paranormal perception of children with SCA as *abiku* (born-to-die-prematurely), there is a need to start rethinking the scope of the responsibilities of caregivers to such children. This would ideally lead to identification of the best way in which to integrate modern medical technology with the valuable culturally sensitive Yoruba approach. The call for a synthesis of traditional and modern outlook does not mean that traditional care practices are inferior to or superseded by Western biotechnological options. There are commendable elements of the Yoruba care for severely genetically impaired children that are useful for the contemporary world.

However, there are also some anachronistic aspects of the culturally sensitive approach that should be reviewed in light of relevant information. For instance, the responsibilities of caregivers in contemporary Yoruba culture should go beyond the boundaries of herbal treatment and spiritual diagnosis. A lack of scientific understanding is perhaps responsible for the fact that the condition is mystified as born-to-die-prematurely syndrome. In jettisoning this popular belief in Yoruba culture, a scientific appreciation of prenatal screening and diagnosis of chromosomal disorders in general is essential.

Uptake of genetic testing and counselling for individuals/couples with known or suspected risk of having a child with a serious genetic disorder is still low among the Yoruba. This new culture can grow when family caregivers are more proactive in encouraging couples with genetic risk profiles to make informed reproductive decisions. Some of the options open to them in this regard include choosing reproductive techniques such as preimplantation or prenatal diagnosis, or reproduction with donor gametes. The importance attached to the capacity for procreation in Yoruba culture is not an end in itself; rather it is a means of promoting the transmission of cultural heritage. This perhaps explains why child fostering practices are commonplace in Yoruba culture, especially among infertile couples.

A critical problem in the Yoruba caregivers' practice and understanding of responsibility for children in general is that it entails only a one-sided conception of being accountable for children's care. The responsibility of caring for children is not strictly about what is in the child's interest. Such intervention only sees the primary care of the child from the perspective of the caregivers in order to instil moral authority and obedience to social conventions and structures. What this sense of being accountable for children's care omits is care for the children's own best interests in terms of developing a sense of autonomy and moral agency. Responsible promotion of children's best interests, especially in the case of children with genetic disorders, requires adequate genetic counselling, diagnosis and a care management plan of SCA. The intention of 'being there for the other,' which is central to caregivers' responsibilities to children in Yoruba culture is laudable, but it is not sufficient. In the words of one author, "[g]ood care should be understood as the practice of combining activities, attitudes and knowledge of the situation" (Gastmans 2006).

13.5 Implications of Family Caregivers' Responsibilities

The suggestion of adjusting the common understanding of responsibilities for children in Yoruba culture has the potential to conflict with some ancient Yoruba cultural values and beliefs, such as the belief in destiny. The Yoruba belief in predestination as typified by *ori* is unwavering, and it suggests that human existence is purposefully designed at conception to fulfil certain ends that are unalterable. Such a belief in destiny has been rationalized as psychologically pacifying in avoiding anxiety about human helplessness in certain situations (Gbadegesin 1984).

SCA is readily explained and accepted as the destiny that cannot be prevented or cured. A genetic reproductive intervention such as PGD may be interpreted as inconsistent with the ontological belief in the destiny of the child. Against this belief, the use of PGD and embryo selection is an efficient way in which carrier prospective parents can avoid passing on the condition while at the same time avoiding an abortion (Okere et al. 2014). Such a choice can prevent the suffering of future children. It can also reduce societal needs for devotion of extra time, cost, care and psychological apprehension that caring for a child with SCA can bring. While embryo selection raises major moral concerns, within the Yoruba cultural matrix it is normatively unproblematic, but does raise an ontological difficulty. Among the Yoruba, there are normative and ontological dimensions to personhood. The moral status of the embryo is determined in terms of its potential to form a part of a caring communal relationship both as a subject and object of solidarity. In principle, the moral status of the embryo increases with its development; however for the Yoruba, it is limited. The reason for this is the embryo's incapacity to coordinate benefiting actions to a network of relationships and its inability to undertake actions for the sake of others.

For the Yoruba, the ontological conception is, however, independent of the normative explanation of personhood. In this conception, human beings are thought to be a conjunction of three interrelated elements: *ara* (body), *emi* (spirit) and *ori* (destiny). In this ontological framework, from the time of conception, a being has these tripartite components in the metaphysical realm. The different states of biological development of a being are only the physical manifestations of her metaphysical existence. As long as this belief in the ontological dimension of being is still held in strong reverence in contemporary Yoruba societies, it has the tendency to obstruct the possibility to make informed reproductive decisions. Critically, however, it should be noted that "cultural practices are not self-justifying; they are justified by appeal to their significance for a moral value" (Gbadegesin 2013: 7). In order to promote the wellbeing of existing children with SCA as well as that of those that are to be conceived, an integration of valuable aspects of caregivers' responsibilities among traditional Yoruba with the knowledge and possibilities brought about by advancements in reproductive technology may be revealing for the contemporary Yoruba world and human society in general. It is important to avail the diagnostic and therapeutic opportunities evinced by advances in genetics technology and genomics without necessarily losing sight of the relevance of the traditional outlook. Family caregivers' responsibilities and support for children is an aspect that has been diminished in many contemporary societies. A core value among the Yoruba is solidarity. However, it is not enough to adhere strictly only to the value of solidarity at the neglect of modern findings in medicine and genetics. Integrating the traditional values of childcare with scientific elements such as preconception care, genetic counselling, diagnosis and treatment of inheritable disorders may better promote the immediate and long term interests of children.

Of course, some of these advancements may raise a number of ethical questions in the local healthcare system. A sustained public discourse is likely to be instrumental in allaying fears and abuses. Developing societies have an obligation to include genetic diagnosis and treatment of seriously genetically disordered children in health care provision. This conveys in the long run a benefit to the sustenance of human wellbeing.

One basic limitation of the constructed responsibilities of family caregivers in contemporary Yoruba culture in an era of genetics/genomics is that such responsibilities may inhibit autonomous reproductive and care choices because of the sometimes over-bearing influence of the family in matters of reproduction. This seems to be the case because the family and communal structure in Yoruba culture are usually predominant over individual autonomous decision. Family caregivers are usually there to care for children with serious genetic disorders. The logic of communality appears countervailing to the logic of genetic counselling and genetic testing which aims, among other things, towards increasing autonomous reproductive decision making. There is therefore an important need to reconcile this tension in the quest to integrate modern reproductive medicine with Yoruba culture.

13.6 Conclusion

A core thread of this chapter is that it is important to rethink the responsibilities of family caregivers to sickle cell anaemia children and adults among the Yoruba in an age of genetic technology. Such a rethinking is not one of total abandonment of one in favour of the other; it is about envisioning the possibility of integration of the traditional and modern outlooks and of preserving the most worthy qualities of each. Beyond the traditional empathetic care for children generally, this chapter argues for a knowledge-emphatic-based care, which entails a synthesis of family caregivers' responsibilities with genetic counselling, diagnosis and management plan of genetic disorders. The integration of modern medicine and genetics into Yoruba culture will involve not only an embrace of contemporary developments, but also a conscious promotion of the value of solidarity, which is the basis of family caregivers' responsibilities.

Cognizant of some imminent practical problems and cultural challenges the newly advocated responsibilities may invite, this chapter argues that the evolution of caregivers' responsibilities attenuates the Yoruba cultural belief in destiny and is an infraction on collective reproductive decision-making in Yoruba culture.

Notwithstanding these potential critical conflicts with the long-standing Yoruba cultural values, this chapter concludes with the recommendation that an ethically informed policy framework in the Nigerian healthcare system should aim for an integration of the traditional childcare outlook with findings and possibilities brought about by contemporary medicine and genetics. An attempt in the formulation of such a framework is the subject of a different work.

References

Ameh, S.J., F.D. Tarfa, and B.U. Ebeshi. (2012). Traditional herbal management of sickle cell Anaemia: Lessons from Nigeria, *Anemia.* doi:10.1155/2012/607436.

Balogun, O.A. 2010. Authentic fatherhood: A traditional understanding. In *Fatherhood: The Dao of daddy*, ed. Lon S. Nea and Micheal W. Austin. West Sussex: Wiley-Blackwell.

Balogun, A.O. 2012. Authentic motherhood: Traditional Yoruba-African perspective. *Φιλοσοφία: International Journal of Philosophy* 41(2): 113–123.

Fayemi, A.K. 2014. Diagnosis of autism, abortion and the ethics of childcare in Yoruba culture. *Indian Journal of Medical Ethics* 11(4): 245–248.

Gastmans, C. 2006. The care perspective in health care ethics. In *Essentials of teaching and learning in nursing ethics: Perspectives and methods*, ed. A. Davis, V. Tschudin, and L. de Raeve. Edinburgh: Elsevier.

Gbadegesin, S. 1984. Destiny, personality and the ultimate reality of human existence. *Ultimate Reality and Meaning* 7(3): 173–188.

Gbadegesin, S. 2013. "Bioethics and an African value system" Keynote address: Teaching skills in international research ethics (TaSkR) V Workshop, April 17 (Indiana University Center for Bioethics), available at https://scholarworks.iupui.edu/. Last accessed June 2015.

Mbiti, J.S. 1970. *African religions and philosophies*. New York: Anchor Books.

Ohaeri, J.U., and W.A. Shokunbi. 2002. Psychosocial burden of sickle cell disease on caregivers in a Nigerian setting. *Journal of National Medical Association* 94(12): 1058–1070.

Okere, C., K. Ailoje-Ibru, K. Olukoya, R. Ogbeche, A. Adewusi, E. Iloabachie, and O. Ashiru. 2014. Successful pregnancy outcome after in vitro fertilisation following Pre-implantation Genetic Diagnosis/Polymerase Chain Reaction screening for single gene disorder (sickle cell anaemia) before embryo transfer: The clinical experience of an in vitro fertilisation clinic in Nigeria. *Nigerian Medical Journal* 55(1): 87–90.

Osibogun, A. 2014. Osibogun: In eight years, We repositioned LUTH. *This day live*, September 28. Available at www.thisdaylive.com/articles/osibogun-in-eight-years-we-repositioned-luth/190080/. Last accessed June 2015.

Taiwo, I.A., O.A. Oloyede, and A.O. Dosumu. 2011. Frequency of sickle cell genotype among the Yoruba in Lagos: Implications for the level of awareness and genetic counseling for sickle cell disease in Nigeria. *Journal of Community Genetics* 2(1): 13–18.

World Health Organization. 2006. Sickle-cell anaemia. Available at: http://apps.who.int/gb/archive/pdf_files/WHA59/A59_9-en.pdf. Last accessed July 2015.

Chapter 14
Parental Autonomy and Responsibility in the Context of Prenatal Diagnosis. Views and Attitudes of Belgian Healthcare Professionals and Families

Sylvia Hübel, Adelheid Rigo, Kathleen Emmery, and Hans Van Crombrugge

Abstract With the expansion of prenatal testing options, we have been challenged to refine our understanding of responsible parenthood. Some argue that parents have a moral obligation to inform themselves about the health status of their future child and select children with the best genetic prospects, while others believe that these steps far exceed parental duty. Much of the public and scholarly attention is directed to the social anxiety surrounding the limits of parental autonomy and the increasing tension between individual rights and community interests. In some instances, women who forego prenatal testing and bear children with certain genetic conditions are deemed irresponsible. There is further an emerging conflict between the policies that benefit individuals and their families and those that benefit the community. Research aimed at investigating these ideas and tendencies has been to date very scarce in Belgium. To gain insight into the views and attitudes regarding prenatal testing and parental autonomy, we conducted in-depth, semi-structured, face-to-face interviews (n = 41) with healthcare professionals involved in prenatal counselling. An online survey was completed by parents (n = 252) with recent experience in prenatal diagnosis.

Keywords Prenatal testing • Diagnosis • Screening • Genetic counselling • Non-directivity • Ethical dilemmas • Parental autonomy

S. Hübel (✉) • K. Emmery
Knowledge Center, Higher Institute for Family Studies, University College Odisee, Brussels, Belgium
e-mail: sylviahubel@gmail.com

A. Rigo • H. Van Crombrugge
Higher Institute for Family Studies, University College Odisee, Brussels, Belgium

K. Hens et al. (eds.), *Parental Responsibility in the Context of Neuroscience and Genetics*, International Library of Ethics, Law, and the New Medicine 69,
DOI 10.1007/978-3-319-42834-5_14

14.1 Introduction

Within the category of prenatal testing (PNT), it is important to make a clear distinction between screening and diagnostic tests. In this article, by *prenatal screening* (PNS) we mean non-invasive methods such as *the combination test* or *the triple test.* These do not involve risks for the pregnancy, but neither do they provide certainty about the presence/ absence of congenital conditions. Prenatal screening only assesses the risk for certain congenital conditions. However, when a screening test indicates an increased risk for an anomaly, parents can choose to have a diagnostic test (such as *chorionic villus sampling* or *amniocentesis*) to confirm the presence or absence of the condition. *Prenatal diagnosis* (PND) is also performed in the case of an increased risk for such conditions, due to maternal age, family history or a previous child with a congenital abnormality. The invasive methods have the advantage of ensuring a conclusive diagnosis, but they give rise to a risk of miscarriage of 0.5–1 %.

With the expansion of these prenatal testing possibilities, the discourse of risk has infiltrated the everyday language of pregnancy, and it has become a central concept in prenatal care. The expectations of 'responsible pregnant behaviour' have been gradually stretched from lifestyle and dietary recommendations to acquiring information on the health status of the unborn and acting upon it. For instance, if we look at the European prenatal screening policies for detecting chromosomal abnormalities, the cut-off of maternal age has been gradually lowered in many countries from 40 to 35 years; in some places, the offer of screening has even been extended to all pregnant women regardless of their age (EUROCAT 2010).

It goes without saying that parental decisions around PNT are not taken in a social vacuum. A whole series of public ideas act to shape this process and a new discourse on parental responsibility is currently being developed. Some argue that parents have a moral obligation to inform themselves about the health status of their future child and select children with the best genetic prospects, while others consider these steps as not part of parental duty (see Herissone-Kelly in this volume). In these polarized discussions, we have on the one hand parents' right to an autonomous decision, on the other, we find a whole array of societal views, based on ideas about normalcy, health, quality of life or disability. Hence, a significant part of public and scholarly attention is today directed at the social anxiety around the limits of parental autonomy and the increasing tension between individual rights and community interests.

There is an emerging conflict between the genetic policies that benefit individuals or families and those that benefit the community. The allocation of healthcare resources, the balancing of the screening costs against the sources needed for the support of the disabled, are the topic of policy discussions. The options created by PNT raise important ethical issues related to the values of solidarity and diversity, but also those of intergenerational responsibility and fair allocation of resources. In some instances, society has come to view women who forego prenatal testing and bear children with certain genetic conditions as irresponsible; sometimes they are even confronted with the accusation that they burden society (Marteau and Drake 1995). Furthermore, there is a growing concern, especially in the US, about the fact that insurance companies could refuse in the future to provide medical coverage for an affected child if the parents knew about the condition in advance. Private insurance companies had already tried to withhold reimbursement for the medical care of children who were diagnosed before birth and whose parents declined the termination of the pregnancy (Holtzman 1998). PNT thus poses new challenges concerning the right to parental privacy and medical confidentiality.

Research focusing on Belgian views and attitudes regarding PNT has been to date very scarce. In a cross-country study, Belgian health care professionals (HCPs) were found "notably" more pessimistic in their counselling about fetal conditions than other colleagues from the region (Wertz and Fletcher 2004). However, this study examined the genetic counselling taking place in specialized, tertiary centres. To date very little is known about Belgian primary and secondary healthcare settings. That is even more interesting to follow up, as international studies found a significant relationship between the views of HCPs and their counselling. For instance, a statistical relationship has been revealed between personal views on abortion and the direction of counselling (Wertz and Fletcher 2004).

Previous local studies mostly looked at the basic organisation of prenatal testing services and the lack of regulations in this regard (Vamos et al. 1997; Helsen et al. 2009). The only large-scale empirical study carried out by Helsen et al. (2009), reported on the practice of prenatal screening for Down syndrome and neural tube defects in the regions of Brussels and Flanders. Another study by Van Kelst et al. (2013) explored midwives' views on maternity care.

To date, research directed at the views and attitudes of health professionals regarding PNT is scarce. Hence, we decided to study these uncharted aspects, with a focus on the extent to which the principles of non-directivity and respect for parental autonomy are fostered in the counselling.

14.2 Research Method

In the first part of our study, we aimed to get an insight in HCPs' views and attitudes regarding prenatal testing. Also, we wanted to assess their practices, with a focus on their counselling style and respect for parental choice. In-depth, semi-structured, face-to-face interviews (n = 41) were conducted with general practitioners (GPs), midwives and gynaecologists involved in prenatal counselling. To ensure a breadth of responses, we used data triangulation through the geographical situation, institutional and person selection. We recruited HCPs from distinct parts of Flanders and the region of Brussels; we contacted different types of healthcare institutions (private practices and small hospitals as well as university centres); the invitations for participation aimed to reach the three professional categories in equal ratios. The interviews lasted between 25 and 70 min. We used a grounded theory thematic approach to analyse the data (Strauss and Corbin 1998); the initial coding was conducted by the first author (SH), and two of the co-authors (AR & KE) cross-checked the identified codes for consistency.

In the second part, our aim was to provide a general impression of how parents think about the available screening and diagnostic options, to explore the factors influencing their choices and to see if they experienced any constraints in the process. Due to the limited timeframe and human resources of this project, in this part of the study we used an online survey, as this permitted us to engage quickly with a greater number of parents. 252 online questionnaires were completed by parents with a recent pregnancy experience (less than 3 years). The survey tackled a broad range of issues such as information delivery, the process of decision-making regarding the available prenatal tests and the overall evaluation of parental experiences and expectations. Most of the questions were multiple-choice. However, we provided space through some open-ended questions for reporting on issues that did not come up in the survey. We used the Qualtrics software for handling the results of the study.

14.3 The Local Practice of Prenatal Screening and Counselling. HCPs' Views

In Belgium, counselling on prenatal testing possibilities is a task shared by general practitioners (GPs), gynaecologists and midwives. Most women see a GP for confirmation of the pregnancy and after this appointment, they are referred to a hospital to undergo their first ultrasound check-up before the end of the first trimester. According to the GPs participating in our study, women rarely return to them, and further care is mostly provided by gynaecologists. All professional categories noted that a small but gradually increasing group of pregnant women chose to be followed-up by midwives active in private practices. They also underlined that these women had very clear views and expectations about the pregnancy care they desired. For instance, they chose midwives in an explicit attempt to minimise medical and technological intervention in the pregnancy, or if they were preparing for an elective home-delivery. The HCPs emphasized that there were no formal agreements or directives regarding the distribution of tasks. This was indicated by most of them as a problematic point because this unstructured system didnot ensure clear communication on the information delivered by each of them.

HCPs pointed out that only a very small group of parents rejected PNS. Though it was difficult to estimate the total number, they placed it somewhere around 5–10 % of all those who have been offered these interventions. This fact has been confirmed by the figures resulting from our survey on parental experience (4 %). HCPs considered that those who rejected PNS were people who made well-reflected, very conscious choices consistent with their values and beliefs. For example, they preferred low-technological pregnancy care. In certain ethnic groups, the great majority of prospective parents rejected all offers of prenatal tests due to their religious convictions precluding abortion (e.g. the Muslim or Jewish populations).

A larger scale Dutch study also confirmed lower uptake rates in these populations (Slagboom 2011). Nevertheless, others showed correlation not only with religious views but also with factors such as inadequate information processing, language barriers or gender constraints (Dormandy et al. 2005; Fransen et al. 2011).

14.4 Good Prenatal Genetic Counselling. Professional Views and Parental Experiences

The HCPs indicated that the goal of prenatal genetic counselling is to help parents reach a well-informed and considered decision, in agreement with their personal values and beliefs. According to them, the main points of counselling included: the presentation of prenatal testing possibilities, the concrete description of the procedures, and the discussion of the risks as well as the notions of false positives/negatives.

First of all, explaining as objectively as possible the tests, both the screening as well as the invasive techniques. After that, listening to people's questions. And eventually, reflecting further with them on the possibilities. (midwife)

It is my task to explain these things to them, to answer their questions and to guide them. I support them whatever their decision might be, but the decision is always theirs. It would not be OK if I told them to do this or that. (midwife)

Many of them dwelled on the importance of clarifying the difference between screening and testing results.

We need to explain the difference between a screening test and a diagnostic test, and this is not always obvious. It is very important to emphasize that screening does not offer a final answer, it is just a risk estimation. (GP)

Eliciting parents' personal views, beliefs and life plans were also considered crucial steps in the counselling process. HCPs emphasized the need to help parents consider all their options and to encourage them to follow their own values, intuitions and feelings so that they make the choice they can best live with. Previous studies had also recommended inquiring into counselees' moral stances to facilitate an informed choice consistent with their personal values (Farrelly et al. 2012; Anderson 1999).

As a professional, you have to bring up all your skills, let people communicate, let them reflect on things. There are no easy solutions. They need to discover the complexity of these issues. First of all, they need to be well-informed, and secondly, they need to listen to their own feelings and follow their own views. It is important that they do not feel pushed to do something. We are trying to place everything in a larger perspective. One of the tricks that helps is, for instance, the "what if?" question. (GP)

HCPs emphasized the necessity to think further with the patients about the implications of the test results for their individual lives. Reflecting on the possible outcomes before testing was also perceived as part of the coping process in case of positive results. For instance, to assess parents' coping capacities, HCPs initiated discussions about their ability and willingness to raise a child with a disability and about their views on pregnancy termination. They elaborated extensively on the "uselessness" of PNS if there was no parental wish to act upon a positive result with further diagnostic tests. They considered that PNS would just cause anxiety and burden the pregnancy experience in such cases.

I think it is important that if you offer a test they should know its implications, otherwise it can be a roller-coaster. (gynaecologist)

The online survey indicated that the vast majority of parents were informed about the PNT options by gynaecologists (92 %), with only a smaller group having received information from GPs (25 %) or midwives (26 %). The parents were mostly satisfied with the information provision, reporting that HCPs clarified and discussed with them the aims of the screening tests (71 %). The majority (82 %) reported being aware of the possible decisions triggered by a positive PNS result and having reflected on these in advance. Concerning PND, a large majority of the parents (87 %) reported having discussed the content, the aims and the risks of the procedures with the HCP.

14.5 On the Feasibility and Desirability of Non-directive Counselling

There is a strong consensus in clinical guidelines all over the world that healthcare professionals (HCPs) active in prenatal genetic counselling should provide prospective parents with adequate information and support so that they can make autonomous, informed decisions in agreement with their personal values. However, empirical studies indicate that the way in which information is provided, as well as counsellors' views and attitudes, can influence parental decision-making (Wertz and Fletcher 2004). Even though non-directivity has been very long valued as the primary guiding principle in genetic counselling, concerns have been raised around its feasibility (Weil 2003).

HCPs all expressed an awareness of their professional task to stay non-directive, and the majority saw this goal as achievable, even if not in absolute terms. They noted a certain tension between the ideal of non-directivity and its achievement in practice. They admitted that even though they were striving for a non-directive approach, they did not always succeed in it. The main reasons quoted as impediments to absolute non-directivity were: *personal background* – HCPs are not 'tabula rasa', they consciously/unconsciously bear the traces of their personal context, value system and education; *nonverbal communication* – unconscious messages hidden in the tone or the body language of the HCP could, for instance, reveal their views. By these statements, they confirmed the findings of earlier studies demonstrating that counselees could read between the lines, picking up cues from healthcare providers' approval or disapproval of their choices (Weil 2003;

Anderson 1999). Some of our HCPs considered that even the offer of a particular PNT method could in itself direct the patient.

Concerns have also been raised in the literature regarding the desirability of non-directivity. Critics have argued that nondirective counselling is insufficient and inadequate under specific circumstances (van Zuuren 1997). A growing number of studies claim that non-directivity does not always answer the needs of counselees and that it could be perceived as abandoning them to some extent or not living up to HCPs' professional responsibility (Hodgson and Spriggs 2005; Chieng et al. 2011). Moreover, some practitioners considered that the ideal of non-directivity was incompatible with their notion of a caring doctor-patient relationship. Instead, they emphasized the need for facilitating informed choice and promoting the emotional well-being of the counselees. More recent research confirms the fact that due to the ever-growing complexity of PNT, counselling on these issues needs more flexibility and creativity. It indicates a trend of moving away from the abstract notion of non-directivity and shifting the focus to creative practices of psycho-social support and shared decision-making (Schwennesen and Koch 2012).

Our online survey showed that prospective parents expected, first of all, adequate information about their possibilities, without any coercion. However, a small minority (7 %) wanted the active involvement of the HCP in the decision, and 25 % of them expressed the wish for more specific, direct advice or the self-disclosure of the HCP on what she would do in their place. They explicitly demanded more directive support in situations where they were under time pressure or where they felt overwhelmed by the difficulty of choices. HCPs pointed out the fact that besides unbiased information, parents also desired an empathetic relationship with the HCP and guidance in their decision.

Such expectations and demands were confirmed by previous studies as well. For instance, the empirical findings of a European study (Kovacs 2011) revealed that even though the majority of women preferred non-directive counselling, some explicitly demanded more directive support. The high occurrence of requests for the disclosure of counsellors' views was also found to be common in other studies (Peters et al. 2004; Thomas et al. 2006). Though, it is not clear whether and to what extent such requests would influence the decision-making process itself or the outcome (Paine et al. 2010; Redlinger-Grosse et al. 2013).

The HCPs regarded the parental demand for disclosure of the counsellor's views as understandable, as prospective parents were lay people looking for validation and decision support. Further, they expressed the awareness that prospective parents sometimes wished to share or avoid the responsibility of choice:

> *May I make a comparison? If I have to see a solicitor or an accountant and I have no clue about their field, but I wish to do what an expert considers the best option, I also often ask those people: 'What would you do in my place ?' If they try to give me an objective answer, I still would ask: What would you do, were you in my shoes? (...) I do not have the experience, and I do not consider it inappropriate to ask now and then, and let myself be guided by someone who knows something about the issue. This can often be reassuring, giving a calm feeling... letting yourself be led by someone with experience is not always wrong. (gynaecologist)*

Table 14.1 HCPs' views on the feasibility and desirability of non-directive counselling

1. **Feasibility:**	
Inadvertent directiveness – the values of the counsellors show in spite of their efforts to remain neutral (see Weil 2003)	*I try to explain all the options as objectively as possible, that is not so easy because as an HCP you have your own vision and opinion … but I try to let that go ... My task is to provide them with objective information, explanation and support in whatever choice they make. (midwife)*
	Probably you always direct a bit, consciously or unconsciously. I think you cannot entirely switch off your own person in a patient-doctor relationship, but I hope it is possible. It probably never works out in absolute terms. I think you still let your personal ideal have a role somewhere, but hopefully, as little as possible. (gynaecologist)
	It is our task as HCPs to let people be themselves and show understanding for all situations. We are trained for this and we need to strive for it. (midwife)
	I'm not sure that non-directivity is 100 % achievable. Your verbal and nonverbal communication, the way you sit and talk … As an HCP you still have your own values and norms. (GP)
Inevitable directiveness – resulting from the inevitable choice of a counselling method and the information provided	*For instance, if, in a case of a very severe abnormality, I say 'this is a serious disability' am I being directive or not? What if the parents had mentioned before that in case of a very severe disability they would terminate the pregnancy, and in case it was not that severe, they would not? So I give objective information, but it is directive. Because I need to tell the truth and the truth is sometimes a bit directive, but it is still up to them to decide. (gynaecologist)*
2. **Desirability:**	
Parental demand for HCPs' self-disclosure	*It happens quite often that they ask my personal ideas. I try then to review with them the advantages and disadvantages, but I never give my personal opinion. (GP)*
	I always say that I actually cannot answer that question. I am not the one who has to raise a child with Down, but they might have to. People ask that, they wish to know what you would do? I never answer it. I always say that "it is such a personal issue, it depends on your coping capacities, on your situation, the number of children you already have, your age and first and foremost it depends on your opinions as a couple around it"…. I am very careful in giving my personal vision. (gynaecologist)
	Then I tell them that my situation is probably not comparable to theirs. I try to avoid this. We are not supposed to give our own opinion during counselling. (GP)

Moreover, HCPs noted that some parents did not have the necessary coping capacities, and they preferred a paternalistic approach. Nevertheless, the great majority of HCPs never shared their personal opinion with the parents. They considered staying neutral and withholding their own views to be the correct professional stance. They mostly deflected the question, emphasizing that people should answer from their unique life context (see Table 14.1).

Only a small group of HCPs shared their personal opinions and gave direct advice on what they would choose in a given situation, but they still emphasized that the final decision belonged to the counselees. They also reported cases where more directive interference was needed, for instance, in the case of counselees with an intellectual disability, or in critical situations under time pressure. Consistently with earlier studies, HCPs showed increased self-disclosure the more they were advanced in their career. Earlier research associated this attitude with accrued experience and increased confidence when deviating from professional standards (Balcom et al. 2013).

14.6 Routinization of Prenatal Testing and Respect for Parental Autonomy

Even though prenatal tests were designed to enhance parental choice, prospective parents can be driven on a path of "choiceless choices" in this context. Intra-uterine therapeutic options are as yet very scarce, therefore, in the case of a conspicuous test result, prospective parents can be confronted with ethical dilemmas regarding their responsibility for the future child, the impact of life with a disabled child on the children they already have, or on their extended family. They report feeling overwhelmed when faced with such situations and trapped in a scenario where no matter what they decide, negative consequences will follow (Leuzinger-Bohleber et al. 2008; Fischmann and Hildt 2011). Besides this lack of 'real choices', the exercise of parental autonomy can be restricted in many other ways, ranging from incomplete or biased information provision to implicit or explicit forms of pressure.

The decision of whether or not to undergo prenatal screening is in itself complex. *What influences prospective parents to accept prenatal screening and testing?* Some critics argue that the mere existence of prenatal diagnostic tests places a moral responsibility on prospective parents to comply with them (Hildt 2002). To assess whether PNT is perceived as a moral imperative, we investigated in our questionnaire what underlying motives prompted expecting couples parents to opt for it. We found that prospective parents choose PNT as they wish to do all they can for their future child. They claimed that considerations concerning the quality of life of the child and the avoidance of suffering predominated in their choices. The majority were looking for reassurance and confirmation of a healthy baby.

Numerous studies have noted that prospective parents' decisions in favour of prenatal tests are also shaped by the way in which HCPs present the information. First of all, the way information is presented can obscure the optional nature of tests. For instance, the routine offer of PNT on a large scale undermines parents' autonomous choices. Prospective parents may think that whatever is offered must be the recommended path. Surveys indicated that they found it difficult to refrain from tests: once they were offered, they felt obligated to have them done (Schmitz et al. 2009). Hence, this routinization could gradually affect individual and social perceptions of responsible parenthood. And last but not least, the minimization of screening tests, presenting them as harmless, "simple blood tests" or introducing them in ways such as "let's see how your baby is doing" can obscure the decisions triggered by a positive result. A conspicuous screening test can force prospective parents into decisions regarding invasive testing that they would not otherwise have taken.

Our survey revealed that the overwhelming majority of women undergo some form of prenatal screening to assess the health status of their future child. Most of them had ultrasound scans during pregnancy (98 %) or the combination test (94 %). *Ultrasound* is the best-known and most widespread non-invasive technique. It is used for instance to screen developing foetuses for structural defects. *The combination test* is one of the most commonly used forms of prenatal screening performed during the first trimester of pregnancy. It calculates the risk of congenital handicaps (such as heart disease or chromosomal abnormalities) based on an ultrasound measurement of the fluid under the skin at the back of the foetus's neck (nuchal translucency) and a maternal blood test.

A minority of the participants in our study have also undergone invasive tests such as *chorionic villus sampling* (9 %) or *amniocentesis* (16 %). *Chorionic villus sampling (CVS)* is performed between 10 and 12 weeks of gestation, and it tests for chromosomal abnormalities. It involves the insertion of a thin needle through the cervix or the abdominal wall to remove cells from the primitive placenta (chorion) surrounding the foetus. It is a highly accurate method of prenatal diagnosis for certain chromosomal abnormalities such as Down syndrome, but it carries a 1 % risk of complications and pregnancy loss. A*mniocentesis* usually occurs after 15 weeks of gestation. The procedure uses an ultrasound scan for guidance to prevent injuries while a fine needle is inserted into the mother's abdomen and uterus. Cells from the amniotic fluid surrounding the developing foetus are collected to test for chromosomal abnormalities and neural tube defects. The risk of complications is about 0.5 %, and these include bleeding, cramping and pregnancy loss.

The GPs and midwives presented as a point of concern the fact that prospective parents complied too easily with screening policy. They considered that PNS is offered as a routine part of pregnancy care in the clinical setting, and they contended that a significant proportion of these prospective parents do not properly understand the further steps triggered by screening. They explained the high rate of participation as a form of routinized and sometimes ill-informed compliance. Hence, they found it necessary to emphasize in their counselling that parents have also the right not to choose any of the available options (see Table 14.2).

Evidence of routinization and differences between professional groups in the focus of counselling have been revealed by earlier Belgian studies as well. Helsen et al. (2009) indicated that 93 % of prospective mothers were offered the screening tests as a routine practice. Furthermore, it showed significant differences regarding counselling focus between GPs, gynaecologists and midwives. It noted that, while midwives emphasized freedom of choice as the most important issue, general practitioners and gynaecologists regarded medical information as the most significant. Midwives also considered women's lived experience and the ethical

Table 14.2 The routinization of prenatal tests (GPs' and midwives' views)

The screening already happens naturally. Most people do not know that they get a nuchal translucency measurement. "May I make an ultrasound, madam? ... OK, the nuchal translucency is this much" done, without giving further explanations, yes, it is all routine. (midwife)
Every pregnant woman now gets the triple test, it is done by default, and some parents do not even realize what consequences follow if it is positive. (GP)
I have some ethical objections about the standard or routine offer of screening. Why is it necessary to offer it to all young and healthy pregnant women? I also have some issues concerning the costs involved. Does society have to pay for them? And why do we want these tests? Do we want to exterminate the disabled? May disabled children not be born anymore? (midwife)
In the hospital, they do the nuchal translucency often before the patient even understands what the consequences can be. The gynaecologist says "You need to come around week 12 for that test" and then I ask them, do you know what that means? "Yes, they do it to see if there is any abnormality" ... and then I start to nuance the differences: it gives only a risk estimation, etc. It is hard, technology, the hospital influence people quite a lot. Especially in a first pregnancy, it's all new to them, and people want to do it well, they do not want to miss any tests. People usually comply, they rarely say "No, I do not want it" (GP)
I remember cases of parents who came here to our practice, people who had the triple test without clear information about it. They suddenly were confronted with a positive test result, and they were quite surprised. I would not say that they were pushed further towards a chorionic villus sampling or amniocentesis, but they did not see another way out, they were already so anxious. (midwife)
I often see people who easily said yes to one or the other prenatal test, and they come back to me saying: "we had an amniocentesis, but we do not really know if we wanted it. What if the test comes back positive? What should we do about it?" I think they should have known that beforehand. If I do it, this might be the outcome, if not, these can be the consequences. My experience tells me that things are not always well explained in the hospital. (midwife)
Screening seems to be automatic, even without asking approval or having a conversation about it. They get a nuchal translucency measurement and a triple test without even knowing it and without giving their consent. (GP)
Women come here sometimes after having had the whole gamut of screening tests, and they are very surprised when we tell them that they could have chosen not to undergo them. (midwife)

issues more important than the other two categories of healthcare professionals did. Another study confirmed this last aspect, showing that midwives argued for a more woman-centred care and woman-defined bodily knowledge as opposed to medicalized knowledge of the body (Van Kelst et al. 2013).

On the other hand, prospective parents' reports confirmed that some healthcare professionals undertook screening tests without their consent. In 24 % of the cases, it was the gynaecologist who made these decisions. The ultrasound and blood tests were presented as standard practice, without adequate information about their purpose or the ways in which the results could influence the course of their pregnancy. Since treatment for most of the screened disorders is not available, in some cases, the results could force them to take further decisions, or cause anxiety and distress.

A rather large group of prospective parents (28 %) has been confronted with the decision of whether or not to proceed with invasive diagnostic tests after a positive screening test and most of them (64 %) experienced this as a difficult decision due to the risks of a miscarriage.

Further, the fact that others around them know about the pregnancy by the time they receive the results is often perceived by prospective parents as a constraining factor. This implies that in the case of a positive result, parental actions and decisions for/against the termination of the pregnancy could not remain private. This can attract in certain settings criticism or even stigmatization. However, our findings suggest that Belgian prospective parents were not overtly pressed in this regard: only 10 % of them experienced disapproval in their environment regarding their decisions.

HCPs unanimously emphasized that decisions regarding prenatal diagnosis should always be well-informed and based on the autonomous choices of the prospective parents. They considered that every parent should be informed about the available options without being constrained to choose any of them. They regarded the use of PNT as a personal choice and not a parental duty. The majority reported respecting parental choice, even when, for instance, parents chose invasive prenatal tests without any medical indication, provided they were well-informed about the content and consequences of the intervention. An important point that should be added in this regard is that the Belgian healthcare system allows and reimburses practically all prenatal screening and testing choices (except the new non-invasive prenatal tests, which are not reimbursed yet) so HCPs can hardly refuse them. Only a very small minority of HCPs considered that parents did not have the right to demand such procedures without indications (see Table 14.3).

The majority of prospective parents rejected the idea of parental responsibility to select the healthiest possible children. Instead, they underscored the responsibility to respect lifestyle recommendations during pregnancy and to give the best possible care after birth. However, it is noteworthy that, while HCPs unanimously rejected the hypothesis of implementing prenatal testing as mandatory, a quite significant group of the prospective parents (between 20 and 30 %) thought that these should be imposed in certain situations (such as high genetic risk running in the family, maternal age above 37 years or already having an affected child, see Table 14.4). Furthermore, while HCPs unanimously believed that they should respect the choices of the prospective parents, 19 % of the prospective parents themselves considered that HCPs should not always follow parental choices. These findings can be interpreted to some extent as a confirmation of the much-feared tendency towards the public support of mandatory testing. This group of prospective parents questions the personal autonomy of other parents, which may suggest diminished solidarity with the groups at risk of being affected by genetic conditions.

Table 14.3 Respect for parental autonomy and choice (HCPs' views)

Absolute respect for parental choice	*This is an important principle: if people wish to do something other than what you consider an intelligent or good choice, you have to respect it. Otherwise, you would go quickly in the direction of Nazism and so. (...) This is a really important principle. And obliging couples to take certain tests, it makes me shiver, really. I find this a terrifying idea. (gynaecologist)*
Parental request for invasive tests without any medical indication	*I find this a very difficult discussion...I try to explain that their request is beyond the norms, but I cannot forbid it, if they wish I will do it ... At the moment it is all reimbursed, so I find it difficult (...) the system is such that if people come to ask for instance for an amniocentesis, we do it. (gynaecologist)*
	If people insist and say for instance that "it looks all in order, but I really want an amniocentesis, no problem we do an amniocentesis. We live in a country where everything is possible, so we will make then an amniocentesis ... But I don't think that it happens too often. It is less than once per year that someone would ask for an amniocentesis without any reasons, out of mere psychological distress. (gynaecologist)
HCPs reactions to the hypothesis of making prenatal diagnosis mandatory	*Oh, no! We should not oblige them to do anything. I am absolutely against mandatory testing. That is the Brave New World of Huxley. I do not believe that a parent is supposed to have prenatal diagnosis. First of all, because you still have the risk of miscarriage. And secondly, you need to know what you want to do with that information. With a prenatal diagnosis, you can even ruin the pregnancy experience by creating unnecessary anxiety and fear. (GP)*
	No, I would not agree with mandatory testing. There is patient autonomy and the right to self-determination, and I would find it absolutely wrong if people were forced to do something. Where are we heading to then? I would find that scary, I get cold shivers at the mere thought of it. (midwife)

Table 14.4 Prospective parents' views on the hypothesis of mandatory testing

Do you think parents should be obliged to undergo certain tests if....?	*The mother is older than 37 years*	*21.49 %*
	There is a certain genetic risk running in the family	*33.06 %*
	They already have a disabled child	*19.01 %*
	Under no circumstances	*63.64 %*
	Total	*100.00 %*

14.7 Professionally Challenging Situations and Ethical Dilemmas. HCPs' Views

Prenatal genetic counselling can pose a number of ethical and professional challenges to HCPs. Even though they expressed a strong commitment to respecting parental autonomy, HCPs also acknowledged some particularly challenging, dilemmatic situations, when they felt conflicted in their practice. Demands for termination of pregnancy in cases of minor anomalies were identified as sources of professional and personal conflict. The most quoted cases were the demands of pregnancy termination for what the HCPs qualified as minor disabilities (such as cleft lip/palate, Turner syndrome, minor spina bifida, missing limb); or selective reductions without medical reasons (in the case of twins or triplets). When they had ethical objections to the choices of the expecting couples, HCPs consulted their colleagues or referred the case to the ethics committee of their institution or specialized centres.

Furthermore, they mentioned the difficulty of communicating probabilistic information. All HCPs dwelled on the importance of explaining what a statistical calculation of risk meant. They considered this as one of the most challenging aspects of counselling, mainly because parents had difficulties in understanding the difference between screening and testing. Moreover, prospective parents wished for and expected clear information instead of intricate estimations.

Some doctors who worked in hospitals where the great majority of the patients were immigrants reported that language barriers and poor understanding of genetics made the uptake of information difficult. They tried to avoid a paternalistic, directive approach, but they contended that non-directive counselling was not appropriate in these situations. They were deeply concerned about the validity of informed consent when prospective parents showed inadequate understanding. However, some hospitals introduced creative solutions to deal with this problem, by using online translation services. HCPs also expressed their concern about situations where patients were unable to make well-informed, autonomous choices because of intellectual disability, family pressure or drug addiction. Tensions inside the couple and disagreements concerning the desirable course of action were also presented as very challenging (see Table 14.5).

14.8 Limitations of the Study

Our study is restricted in size, and it does not represent a generalizable account of the Belgian situation. Nevertheless, it reaches its heuristic goals of examining emerging trends and through this, it informs a further larger-scale analysis.

Furthermore, we are aware that HCPs' self-reporting can reflect professionally and ethically desirable views and ideal practices rather than everyday attitudes and reality. What people actually do in their counselling can depart from what they say they would do. The assessment of actual counselling could be achieved through large-scale patient reports and observational studies.

Our intention in the parent survey was to reach both prospective mothers and fathers in the recruitment process. However, the number of male respondents was extremely low. Subsequently, we targeted and repeatedly addressed sources where prospective fathers could be better reached, with no significant improvement (97 % women and only 3 % men). Ethical discussions on the use of PNT should be set against the background of a family-framed model of responsibility. Literature confirmed that women take their prenatal decisions with an eye to the whole family.

Table 14.5 Professional challenges and ethical dilemmas

Explaining the statistical calculation of risk	*It is very difficult for them to understand what a screening test is and what low risk or high risk* means. *Lots of people think that when it comes to these tests, you will be able to say it is a Down, or it is not a Down ... It is hard for them to understand that it is only a risk calculation, an estimation and not a diagnosis. So explaining the difference between screening and diagnostic tests is very difficult. (gynaecologist)*
Conflict between their sense of responsibility for the future child and parents' choice	*For instance, we are following up here in the hospital a family that already has two children with a severe kidney problem These children need heavy medication, and they will probably need a kidney transplant as well, and these parents keep having children without using prenatal diagnostic services or PGD, and they will probably have another child with such a dysfunction. I find this very difficult, I find this an ethical issue ... you cannot oblige people ... I personally have difficulties with this, if the technical possibilities exist and they are not being used. (gynaecologist)*
Termination of pregnancy for minor conditions (they referred mostly to cleft lip/palate, Turner syndrome, Klinefelter, deafness, a small spina bifida).	*For instance a cleft lip ... I absolutely don't see a cleft lip as a reason to terminate a pregnancy. We have rejected this in the hospital, and the pregnancy was not yet very advanced so they could still go to the Netherlands... But these are issues where I could say "I should do it, because they will do it somewhere else". I still don't do it. I will follow my own conscience and our norms and values here (hospital). If people do not wish to follow these, and they go somewhere else, it is their responsibility. (gynaecologist)*
	Pregnancy termination for not really major things ... or the other way round, for instance, in the case of an anencephaly, and people say: "we do not want a termination". (gynaecologist)
	My limit is the quality of life of the baby .... Can it have a quality of life in spite of the limitations ...? And secondly: don't we place too big a burden on the parents? When the baby is sent out of the hospital, we are rid of it, we do not have to deal with all the consequences of the decisions, the consequences are for the parents. (gynaecologist)
	Some parents go way too far in wanting a perfect child and at the smallest thing they notice, they ask for an abortion ... we cannot do that. There should be a good reason ... I mean, there should be clear abnormalities or a serious disability so that we stay inside the legal norms, but some parents ask for things that are not really possible. (gynaecologist)
	If someone came to ask me to terminate a pregnancy – and this really has happened after the confirmation of a cleft lip through ultrasound. It was a couple with a very low coping capacity, and they asked for the termination of the pregnancy. I told them "look, I can't do that ... I understand that you fear everything that would follow but I myself cannot do this". Finally,they did not do it...but I told them 'look, if you really want to proceed with the termination I can refer you somewhere else'. (gynaecologist)
Lack of understanding of genetic information because of language barriers, low educational background or intellectual disability	*There is only a small group of people that really understands genetic information. I work in a hospital with a majority of patients of foreign origin. And here comes the ethical concern: the decision on whether to choose for or against prenatal diagnosis should be theirs. However, for an informed choice, they should be able to understand what this is all about. For most of my patients, it is impossible to get to this understanding. I have struggled with this issue already for so long, I think: If I do not proceed with screening, I deny them a service which is available and they have the right to it. The other option would be to proceed anyhow with testing, but that would be paternalistic, and I am not a big fan of this approach. (gynaecologist)*

They are called upon to reconcile several competing interests: commitment to the baby as well as a commitment to preserving the stability of their relationship and the wellbeing of any existing children. However, to date, most studies report on women's accounts of prenatal diagnosis and men's perspectives appear to be underexplored. Previous research also reported low-response rates from men and confirmed that women had been delegated this responsibility inside the couple (Locock and Alexander 2006; Gottfredsdotti et al. 2009). Hence, further studies are needed to understand the ways in which responsibility around prenatal testing is interpreted and negotiated inside families.

Another possibly important limitation in the profile of the responding parents could be their educational background, as the great majority were highly trained (83 % post-secondary studies, university or non-university level). Our respondents to the study may also differ from non-respondents through the fact that they chose to participate out of a particular interest in the topic (such as having a family member affected by a genetic condition). Some of them openly shared this motivation, and those who had been confronted with a positive screening or test result represented 31 % of the respondents.

14.9 Conclusion

In the context of prenatal diagnosis, the idea of parental responsibility has been shifting to areas which had not previously been seen as a matter of parental choice. Hence, the topic is receiving increased interest in public discourse, and the current debates are rather polarized. At one end of the controversy, those echoing consequentialist or communitarian principles argue that parents have a moral obligation to select children with the best genetic perspectives for the sake of the child and/or of the community (for a critical discussion of such an argument, see Herissone-Kelly, Güell, and Hens in this volume). Others argue that the essence of good parenting is accepting, loving and caring for one's children unconditionally, regardless of their particular traits and abilities (Parens and Asch 1999; Sandel 2007).

Our research revealed a cautious professional attitude in agreement with the contemporary emphasis on respect for patients' autonomy. Belgian HCPs involved in counselling around the issue of prenatal diagnosis reported a real preoccupation with promoting prospective parents' autonomous and informed choices. The vast majority felt able to distance themselves from their personal views in morally conflicting situations. However, their accounts also revealed ethically and professionally challenging situations, where they would have preferred clear guidelines. A large majority of the parents were satisfied with prenatal counselling, though some of them desired more emotional support and active involvement of HCPs in their decision-making. They also reported ethically and psychologically challenging situations.

Therefore, acknowledging the difficulties experienced on both sides of the counselling situation can be an excellent starting point in opening up a societal debate on the regulation of prenatal diagnostic practices. Our study provides insight into Belgian HCPs' and prospective parents' views, and can be very helpful in mapping their main concerns and dilemmas. However, the need for further large-scale research, multidisciplinary consultation and broad social debate clearly emerged. PNT is still in a process of very rapid expansion, both by being offered to more and more pregnant women and by being able to detect ever more conditions. In this rapidly changing setting, the practice should be critically re-evaluated, reassessing both prospective parents' and professionals' needs especially with regard to ethically problematic situations.

Author Contribution Statement *Sylvia Hübel* designed the research tools, recruited the participants, conducted the interviews, interpreted the data and wrote the report of the results. *Adelheid Rigo* conceived the project, submitted the research proposal for funding and supported all stages of design and data interpretation. *Kathleen Emmery* participated in the research design and data interpretation. *Hans van Crombrugge* followed up the different steps of the research project and collaborated in the formulation and submission of this chapter.

References

Anderson, G. 1999. Nondirectiveness in prenatal genetics: Patients read between the lines. *Nursing Ethics* 6(2): 126–136.

Balcom, J.R., Veach P. McCarthy, H. Bemmels, K. Redlinger-Grosse, and Bonnie S. LeRoy. 2013. When the topic is you: Genetic counselor responses to prenatal patients' requests for self-disclosure. *Journal of Genetic Counseling* 22: 358–373.

Chieng, W.S., N. Chan, and S.C. Lee. 2011. Non-directive genetic counselling – Respect for autonomy or unprofessional practice? *Annals of the Academy of Medicine, Singapore* 40(1): 36–42.

Dormandy, E., S. Michie, R. Hooper, and T.M. Marteau. 2005. Low uptake of prenatal screening for Down syndrome in minority ethnic groups and socially deprived groups: A reflection of women's attitudes or a failure to facilitate informed choices? *International Journal of Epidemiology* 34(2): 346–352.

EUROCAT. 2010. Special report: Prenatal screening policies in Europe 2010, *EUROCAT Central Registry*. Newtownabbey: University of Ulster.

Farrelly, E., M.K. Cho, L. Erby, D. Roter, A. Stenzel, and K. Ormond. 2012. Genetic counseling for prenatal testing: Where is the discussion about disability? *Journal of Genetic Counseling* 21: 814–824.

Fischmann, T. and E. Hildt. 2011. *Ethical dilemmas in prenatal diagnosis*, ed. New York: Springer.

Fransen, M.P., M.L. Essink-Bot, I. Vogel, J.P. Mackenbach, E.A.P. Steegers, and H.I.J. Wildschut. 2011. Ethnic differences in informed decision-making about prenatal screening for Down's syndrome. *Journal of Epidemiology and Community Health* 64: 262–268.

Gottfredsdotti, H., J. Sandall, and K. Bjornsdottir. 2009. 'This is just what you do when you are pregnant': A qualitative study of prospective parents in Iceland who accept nuchal translucency screening. *Midwifery* 25: 718–719.

Helsen, S., R. Devlieger, N. El Handouni, and V. Van der Stighelen. 2009. Prenatale screening van syndroom van Down en neurale-buisdeffecten: de praktijk vandaag. *Tijdschrift voor vroedvrouwen* 5: 254–261.
Hildt, E. 2002. Autonomy and freedom of choice in prenatal genetic diagnosis. *Medicine, Health Care and Philosophy* 5(1): 65–72.
Hodgson, J., and M. Spriggs. 2005. A practical account of autonomy: Why genetic counselling is especially well suited to the facilitation of informed autonomous decision making. *Journal of Genetic Counseling* 14(2): 89–97.
Holtzman, N.A. 1998. Eugenics and genetic testing. *Science in Context* 11: 397–417.
Kovacs, L. 2011. Taking risk in striving for certainty. Discrepancies in the moral deliberations of counsellors and pregnant women undergoing PND. In *Ethical dilemmas in prenatal diagnosis*, ed. T. Fischmann and E. Hildt, 121–138, New York: Springer.
Leuzinger-Bohleber, M., Eve-Marie Engels, and J. Tsiantis. 2008. *The janus face of prenatal diagnostics: A european study bridging ethics, psychoanalysis and medicine*. London: Karnac.
Locock, L., and J. Alexander. 2006. 'Just a bystander'? Men's place in the process of fetal screening and diagnosis. *Social Science & Medicine* 62: 1349–1359.
Marteau, T., and H. Drake. 1995. Attributions for disability: The influence of genetic screening. *Social Science & Medicine* 40(8): 1127–1132.
Paine, A.L., P. McCarthy Veach, I.M. MacFarlane, B. Thomas, M. Ahrens, and B.S. LeRoy. 2010. "What would you do if you were me?" Effects of counselor self-disclosure versus nondisclosure in a hypothetical genetic counseling session. *Journal of Genetic Counseling* 19(6): 570–584.
Parens, E., and A. Asch. 1999. The disability rights critique of prenatal genetic testing: Reflections and recommendations. *Hastings Center Report* 29: S1–S22.
Peters, E., P. McCarthy Veach, E.E. Ward, and B.S. LeRoy. 2004. Does receiving genetic counseling impact genetic counselor practice? *Journal of Genetic Counseling* 13: 387–402.
Redlinger-Grosse, K., P.M. Veach, and I.M. MacFarlane. 2013. What would you say? Genetic counseling graduate students' and counselors' hypothetical responses to patient requested self-disclosure. *Journal of Genetic Counseling* 22(4): 455–468.
Sandel, M.J. 2007. *The case against perfection. Ethics in the age of genetic engineering*, 45–46. Cambridge, MA: Harvard University Press.
Schwennesen, N., and L. Koch. 2012. Representing and intervening: 'doing' good care in first trimester prenatal knowledge production and decision-making. *Sociology of Health & Illness* 34(2): 283–298.
Schmitz, D., et al. 2009. An offer you can't refuse? Ethical implications of non-invasive prenatal diagnosis. *Nature Reviews Genetics* 10: 515.
Slagboom, M. 2011. *Echo. Prenataal onderzoek en keuzevrijheid*. Amsterdam: Augustus.
Strauss, A., and J. Corbin. 1998. *Basics of qualitative research: Techniques and procedures for developing grounded theory*, 2nd ed. Thousand Oaks: Sage.
Thomas, B.C., P.M. Veach, and B.S. Leroy. 2006. Is self-disclosure part of the genetic counselor's clinical role? *Journal of Genetic Counseling* 15(3): 163–177.
Vamos, E., K. Vandenberghe, and J.J. Cassiman. 1997. Prenatal diagnosis in Belgium. *European Journal of Human Genetics* 5(Suppl 1): 7–13.
Van Kelst, L., B. Spitz, W. Sermeus, and A.M. Thomson. 2013. A hermeneutic phenomenological study of Belgian Midwives' views on ideal and actual maternity care. *Midwifery* 29(1): e9–e17.
van Zuuren, F.J. 1997. The standard of neutrality during genetic counselling: An empirical investigation. *Patient Education and Counseling* 32(1): 69–79.
Weil, J. 2003. Psychosocial genetic counseling in the post-nondirective era: A point of view. *Journal of Genetic Counseling* 12(3): 199–211.
Wertz, D.C., and J.C. Fletcher. 2004. *Genetics and ethics in global perspective*. Boston: Kluwer Academic Publishers.

Author Index

K. Hens et al. (eds.), *Parental Responsibility in the Context of Neuroscience and Genetics*, International Library of Ethics, Law, and the New Medicine 69,
DOI 10.1007/978-3-319-42834-5

W

Y

Z

Subject Index

K. Hens et al. (eds.), *Parental Responsibility in the Context of Neuroscience and Genetics*, International Library of Ethics, Law, and the New Medicine 69, DOI 10.1007/978-3-319-42834-5

MIX
Papier | Fördert gute Waldnutzung
FSC® C083411

Kolibri 360 GmbH
Ferdinand-Jühlke-Straße 7
99095 Erfurt, Deutschland
produktsicherheit@kolibri360.de